Workshop on Natural Language Processing for Medical Conversations 2020

Online
10 July 2020

ISBN: 978-1-7138-1392-7

ACL 2020

Natural Language Processing for Medical Conversations

Proceedings of the Workshop

July 10, 2020
Online

Introduction

Welcome to the ACL 2020 Workshop on NLP for Medical Conversations.

Primary care physicians spend nearly two hours on creating and updating electronic medical/health records (EMR/EHR) for every one hour of direct patient care. Additional administrative and regulatory work has contributed to dissatisfaction, high attrition rates and a burnout rate exceeding 44% among medical practitioners. Recent research has also linked burnout to medical errors, showing doctors who report signs of burnout are twice as likely to have made a medical error. It is imperative to find a solution to minimize causes of such errors, via better tooling and visualization or by providing automated decision support assistants to medical practitioners.

First steps towards introducing automation for clinical documentation have recently emerged. These include approaches from end-to-end clinical documentation to the development of a dialog system with virtual patients for physician training. Commercial products include offerings from large corporations like Microsoft, Nuance, Amazon, and Google to many upcoming startups.

The goal of this workshop is to bring together NLP researchers and medical practitioners, along with experts in machine learning, to discuss the current state-of-the-art approaches, to share their insights and discuss challenges. This is critical in order to bridge existing gaps between research and real-world product deployments, this will further shed light on future directions. This will be a one-day workshop including keynotes, spotlight talks, posters, and panel sessions.

Our call for papers for this inaugural workshop met with a strong response. We received 20 paper submissions, of which we accepted 9 papers with acceptance rate of 45%. Our program covers a broad spectrum of applications and techniques. It was augmented by invited talks from Tanuj Gupta (Cerner), Adam Miner(Stanford), Anitha Kannan(Curai), Steven Bendrick(OHU) and Judy Chang(UPSM).

Organizers:

Parminder Bhatia (Amazon)
Byron Wallace (Northeastern University)
Izhak Shafran (Google)
Mona Diab (Facebook)
Steven Lin (Stanford)
Chaitanya Shivade (Amazon)
Rashmi Gangadharaiah (Amazon)

Program Committee:

Asma Ben Abacha, Busra Celikkaya, Denis Newman-Griffis, Emily Alsentzer, Fabio Rinaldi, Halil Kilicoglu, Jered McInerney, Jungwei Fan, Kevin Small, Kirk Roberts, Kristjan Arumae, Laurent El Shafey, Luyang Kong, Mohammed Khalilia, Nima Pourdamghani, Preethi Raghavan, Qing Sun, Ramakanth Kavuluru, Sarthak Jain, Sarvesh Soni, Sravan Bodapati, Sujan Perera, Timothy Miller, Travis Goodwin, Tristan Naumann, Vivek Kulkarni, Yassine Benajiba, Yi Zhang, Yifan Peng

Invited Speaker:

Tanuj Gupta (Cerner)
Adam Miner(Stanford)
Anitha Kannan(Curai)
Steven Bendrick(OHU)
Judy Chang(UPSM)

Table of Contents

Conference Program

Friday, July10,2020

09:00–10:30   Morning Session I

09:00–09:15   *Welcome and Opening Remarks*

09:15–10:00   *Invited Talk*
Tanuj Gupta

10:00–10:15   *Methods for Extracting Information from Messages from Primary Care Providers to Specialists*
Xiyu Ding, Michael Barnett, Ateev Mehrotra and Timothy Miller

10:15–10:30   *Towards Understanding ASR Error Correction for Medical Conversations*
Anirudh Mani, Shruti Palaskar and Sandeep Konam

11:00–12:30   Morning Session II

11:00–11:45   *Invited Talk*
Anitha Kannan

11:45–12:30   *Invited Talk*
Steven Bendrick

13:30–15:30   **Afternoon Session I**

13:30–14:15   *Invited Talk*
Judy Chang

14:15–14:30   *Studying Challenges in Medical Conversation with Structured Annotation*
Nan Wang, Yan Song and Fei Xia

14:30–14:45   *Generating Medical Reports from Patient-Doctor Conversations Using Sequence-to-Sequence Models*
Seppo Enarvi, Marilisa Amoia, Miguel Del-Agua Teba, Brian Delaney, Frank Diehl, Stefan Hahn, Kristina Harris, Liam McGrath, Yue Pan, Joel Pinto, Luca Rubini, Miguel Ruiz, Gagandeep Singh, Fabian Stemmer, Weiyi Sun, Paul Vozila, Thomas Lin and Ranjani Ramamurthy

14:45–15:00   *Towards an Ontology-based Medication Conversational Agent for PrEP and PEP*
Muhammad Amith, Licong Cui, Kirk Roberts and Cui Tao

15:00–15:15   *Heart Failure Education of African American and Hispanic/Latino Patients: Data Collection and Analysis*
Itika Gupta, Barbara Di Eugenio, Devika Salunke, Andrew Boyd, Paula Allen-Meares, Carolyn Dickens and Olga Garcia

15:15–15:30   *On the Utility of Audiovisual Dialog Technologies and Signal Analytics for Real-time Remote Monitoring of Depression Biomarkers*
Michael Neumann, Oliver Roessler, David Suendermann-Oeft and Vikram Rama-narayanan

16:00–18:00   **Afternoon Session II**

16:00–16:45   *Invited Talk*
Adam Miner

16:45–17:15   ***Lightning Talk***

17:15–17:30   *Robust Prediction of Punctuation and Truecasing for Medical ASR*
Monica Sunkara, Srikanth Ronanki, Kalpit Dixit, Sravan Bodapati and Katrin Kirchhoff

17:30–17:45   *Topic-Based Measures of Conversation for Detecting Mild CognitiveImpairment*
Meysam Asgari, Liu Chen and Hiroko Dodge

17:45–18:00   ***Closing Remarks***

xi

Classifying Electronic Consults for Triage Status and Question Type

Xiyu Ding[1,2] and **Michael L. Barnett**[2,3] and **Ateev Mehrotra**[4] and **Timothy A. Miller**[1,4]

[1]Boston Children's Hospital, Boston, MA
[2]Harvard T.H. Chan School of Public Health, Boston, MA
[3]Brigham and Women's Hospital, Boston, MA
[4]Harvard Medical School, Boston, MA

Abstract

Electronic consult (eConsult) systems allow specialists more flexibility to respond to referrals more efficiently, thereby increasing access in under-resourced healthcare settings like safety net systems. Understanding the usage patterns of eConsult system is an important part of improving specialist efficiency. In this work, we develop and apply classifiers to a dataset of eConsult questions from primary care providers to specialists, classifying the messages for how they were triaged by the specialist office, and the underlying type of clinical question posed by the primary care provider. We show that pre-trained transformer models are strong baselines, with improving performance from domain-specific training and shared representations.

1 Introduction

Electronic consult (eConsult) systems allow primary care providers (PCPs) to send short messages to specialists when they require specialist input. In many cases, a simple exchange of messages precludes the need for a standard in-person referral. eConsult systems decrease wait times for a specialty appointment. (Barnett et al., 2017) An example eConsult question is shown in Figure 1. In general, these questions are much shorter than, say, electronic health record texts. There is a stereotypical structure to these questions, including short histories, descriptions of the current problem, and questions about diagnosis, medication management, procedures, or other issues. When the message is received by a specialist's office, specialist reviewers in that office determine whether the patient needs to be scheduled for a specialist visit or whether the specialist may be able to answer a PCP's question directly without a visit. If a visit needs to be scheduled, the specialists decide whether it is urgent or not (in practice, whether the

<age> year old woman with newly diagnosed dermatomyositis who also has significant dysphagia to solids greater than liquids. She has been started on prednisone and methotrexate. She is originally from <country> and has had no prior colon cancer screening. We would appreciate an evaluation for both upper endoscopy and colonoscopy. Upper endoscopy to evaluate her dysphagia and colonoscopy for malignancy screening (dermatomyositis patients are at increased risk for malignancy)

Figure 1: An example eConsult question

patient goes to the front of the queue). Because these eConsult messages are unstructured, health systems do not know how they are used. Automatically extracting information about the content and response to these questions can help health systems better understand the specialist needs of their PCPs and guide population health management. Accurately classified eConsults can inform decision-making about how to allocate resources for quality improvement, additional specialist investment and medical education to best serve their patient population.

In this work, we use standard support vector machine (SVM)-based baselines and transformer-based pre-trained neural networks (i.e., *BERT models) to classify eConsult questions along two dimensions, focusing on referrals to gastroenterology and liver specialists.

First, we build classifiers that attempt to learn to predict triage status (e.g., urgent or non-urgent) assigned to questions by the specialist reviewer. Our goal is to use the ability (or inability) of classifiers to perform this task to understand the consistency of scheduling decisions across individual clinicians. This addresses a concern that specialist reviewers vary too much in their judgment on whether a visit is urgent or non-urgent. To do this, we performed experiments that compare classifiers trained on triage decisions of single specialist reviewers. The magnitude of inconsistency or unexplainable

Proceedings of the 1st Workshop on NLP for Medical Conversations, pages 1–6
Online, 10, 2020. ©2020 Association for Computational Linguistics
https://doi.org/10.18653/v1/P17

decisions among reviewers would inform whether these systems can consistently work as intended to reduce specialist visits safely and effectively.

Second, we build classifiers for the task of understanding the implicit information need that is the cause for the PCP asking the question – we call this *question type*. We developed an annotation scheme and annotated a sample of questions from across eight years for five question types. We then train and evaluate several classifier models, including standard architectures with problem-specific additions. Our results show that triage status is difficult to learn in general, but even more difficult between reviewers, suggesting inconsistent reviewer decisions may be occurring. When classifying question type, the best-performing models are domain-specific pre-trained transformers, and that jointly training to predict different question types is the most effective technique. Our best result occurs when combining domain-specific vocabularies with multi-task learning, suggesting that there is a synergistic effect between these two augmentations.

2 Background

BERT (Bidirectional Encoder Representations from Transformers), along its variants, have been proven to outperform other contextual embedding (e.g. ELMo (Peters et al., 2018)) or traditional word embedding models (e.g. Word2Vec (Mikolov et al., 2013), GloVe (Pennington et al., 2014), etc.) in a wide variety NLP tasks.

BERT learns contextual embeddings through pre-training on a large unlabeled corpus (including the BooksCorpus (800M words) and English Wikipedia (2,500M words)) via two tasks, a masked language model task and a next sentence prediction task (Devlin et al., 2019).

Domain-specific BERT models have been released, including BioBERT (Lee et al., 2020), which started from a BERT checkpoint and extended pre-training on biomedical journal articles, SciBERT (Beltagy et al., 2019), which is pre-trained from scratch with its own vocabulary, and ClinicalBERT (Alsentzer et al., 2019) which started from BERT checkpoints and extended pre-training using intensive care unit documents from the MIMIC corpus (Johnson et al., 2016). In this work, we use vanilla BERT, SciBERT, and two versions of ClinicalBERT, Bio+Clinical BERT and Bio+Discharge Summary BERT[1].

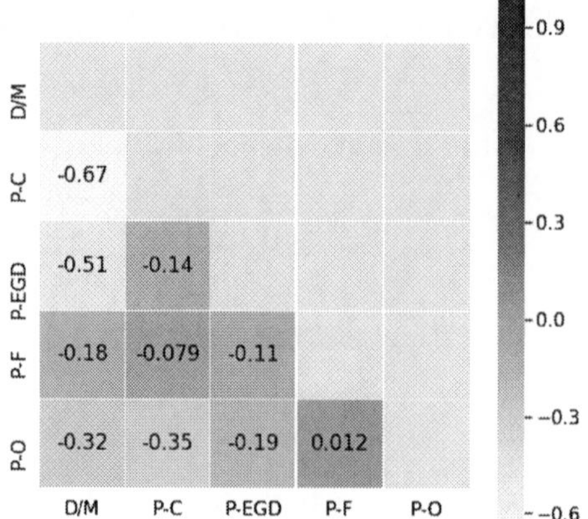

Figure 2: Normalized pointwise mutual information between categories in the question type annotations. Values close to 0 represent variables whose distributions are independent, values above 0 (up to 1) represent pairs of variables that are more likely than chance to occur together, and values below 0 (down to -1) represent pairs of variables that are less likely than chance to occur together.

3 Materials and Methods

3.1 Data

We use de-identified text data from 2008-2017 from the San Francisco Department of Public Health (SFDPH), for which we examined one specialty (gastroenterology and liver) with over 33,000 eConsults.[2] For each eConsult question there are four possible scheduling decisions, *Initially Scheduled* (IS – added to the end of the specialist visit queue), *Not Scheduled* (NS – not added to the queue, typically was resolved via a return message), *Overbook* (OB – added to the front of the queue) and *Scheduled After Review* (SAR – added to the end of the queue after deliberation or additional exchange of messages). Each eConsult also contains meta-data, including a unique identifier referring to the specialist reviewer who first reviewed that question which we use later to train reviewer-specific models. This data was obtained by Boston Children's Hospital under a data use agreement with SFDPH, but unfortunately the terms of that agreement do not allow for public release of the dataset.

For the question type annotation, there are five possible types, *Diagnosis/Management*

---

[1] Bio+Clinical BERT and Bio+Discharge Summary BERT are initialized from BioBERT and respectively trained using MIMIC notes from all types and the notes from discharge summary only.

[2] This research was approved by our institution's Institutional Review Board as "Not human subjects research."

(D/M), *Procedure-EGD*[3] (P-EGD), *Procedure-Colonoscopy* (P-C), *Procedure-Other* (P-Other), and *Procedure-FlexSig* (P-F). These types are *not* mutually exclusive – a question could, for example, request both a colonoscopy and an EGD. Figure 2 shows the normalized pointwise mutual information between each of the five question type categories. For that reason, it should be modeled as multi-label classification tasks, rather than a multi-class classification task.

This set of categories was created by an iterative process, where clinical experts coded samples of a few dozen eConsult questions at a time, refining after each iteration, with the goal of striking a balance between informativeness to leaders at these health care systems, learnability, and ease of annotation. The annotator who performed the question type annotations is a certified medical coder with several decades of experience in coding clinical documents for various administrative and NLP-related categories. We double annotated a portion of the data and scored them for inter-annotator agreement using Cohen's Kappa. Agreements 0.76 for D/M, 0.94 for P-C, 0.87 for P-EGD, and 0.29 for P-O. P-O is difficult to annotate reliably because it is not clear when it needs to be annotated at all – it is a bit of a default category that probably needs clearer instructions for when it should be annotated.

For the triage classifier, we can use all questions in the data set, because they contain automatic labels for the triage decisions that were made by the specialist reviewers. For the question type classifier, we use a sample of 969 questions annotated by our trained medical coder. We divided the data into training, development, and test splits for training classifiers with an 70%/10%/20% split.

3.2 Machine Learning Algorithms

3.2.1 SVM with Bag Features

The simplest method for classifying text uses an SVM with "bag-of-words" (BoW) features. The text is represented by a "feature vector" $\mathbf{v}$ of size V (i.e., vocabulary size) while the value of ith element, v_i equals to the frequency of the ith word in the vocabulary in the document. A generalization of BoW is "bag-of-n-grams" (BoN). N-grams is a contiguous sequence of n items from a given sample of text. A bag-of-N-grams model has the simplicity of the BoW model, but allows the preservation of more word locality information. In this study, we combine the words and n-grams to create the features. Optimal number of n-grams and the hyper-parameter C of SVM are selected by grid search with 3-fold cross-validation. We performed SVMs with BoN features for both tasks as the baseline reference given that it is surprisingly strong for many tasks in document classification.

One mutation of BoW in the clinical domain is the "bag of CUIs" (BoC). CUIs, or Concept Unique Identifiers map the text spans to medical dictionaries and words with the same medical implications are unified to the same concepts. We use Apache cTAKES (Savova et al., 2010) to extract the medical concepts existing in the text data and apply an SVM on the bag of concepts.

We use the Scikit-Learn implementation of SVMs to implement the training and inference (Pedregosa et al., 2011).

3.2.2 BERT Models

We fine-tune the models (updating the weights of the encoders and classification layer) on our tasks with four different versions of BERT models, BERT (base-uncased), SciBERT (base-uncased), Bio+Clinical BERT (base-cased) and Bio+Discharge Summary BERT (base-cased). For both tasks, we use the last hidden state of the [CLS] token as the aggregate sequence representation. The [CLS] representation is fed into an output layer (softmax for the triage status classifier, sigmoid for the question type classifiers) to get the predicted probability of all labels. All parameters from BERT are fine-tuned by minimizing the overall cross-entropy loss. We use the HuggingFace Transformers library (Wolf et al., 2019) for our BERT implementations.[4] We monitor the training and validation loss for each training epoch and save the model with the highest Macro-F1 score on the validation set before testing on the test split.

3.2.3 Multi-task BERT

For the question task, we also explore a multi-task learning scheme which allows us to jointly fine tune BERT for predicting all the labels with the same model. This forces the fine-tuning process to learn representations that are good for multiple tasks, which can potentially benefit as both regularization and by indirectly sharing information between labels that are known to be correlated. For this model, the same [CLS] representation is fed

---

[3]Esophagogastroduodenoscopy

[4]https://github.com/huggingface/transformers

	IS	NS	OB	SAR	Ave.
SVM	0.64	0.46	0.54	0.17	0.45
BERT	0.62	0.48	0.54	0.21	0.46
SciBERT	0.64	**0.54**	0.53	0.15	0.47
Bio+Clinical BERT	0.64	0.50	**0.55**	0.22	**0.48**
Bio+DS BERT	**0.65**	0.49	0.54	**0.24**	**0.48**

Table 1: F1 scores of SVM and BERT classifiers for predicting scheduling decisions

	R1	R2	R3	R4
R1	**0.46**	0.28	0.21	0.33
R2	0.35	**0.43**	0.27	0.39
R3	0.18	0.19	**0.37**	0.23
R4	0.32	0.36	0.37	**0.49**

Table 2: Macro F1 scores showing performance of BERT fine tuned on one reviewer's labels and tested on another.

into five separate output nodes with the sigmoid activation function to get the predicted probabilities of five binary outcomes. The BERT parameters are fine tuned by minimizing the aggregated binary cross-entropy loss of all labels.

4 Experiments and Results

4.1 Triage Status

For the triage classifier, we first train several classifiers first on the entire eConsult training split, and test it on the development split. Results of the SVM with linear kernel and a few fine-tuned BERT models show that training across all consults results in poor performance (Table 1). As noted in the introduction, one explanation is that specialist reviewers were not consistent relative to each other. We thus examined was whether reviewers distributions over triage statuses were similar. Figure 3 shows a histogram of each reviewer's distributions of decisions – there are large differences in what fraction are labeled urgent (*Overbook* category). In order to further investigate the consistency of these scheduling decisions among different reviewers, we also trained four reviewer-specific models. Table 2 shows the results of each reviewer-specific model on text from other reviewers. Column headers indicate the reviewer used to train the model and rows indicate test reviewer.

4.2 Question Type Classification

We evaluated several different architectures on this task to explore the value of domain-specific information, as well as the importance of sharing information between the different labels. Tables 3 and 4 shows the results of the experiments for the question type classifiers. We omit results for P-

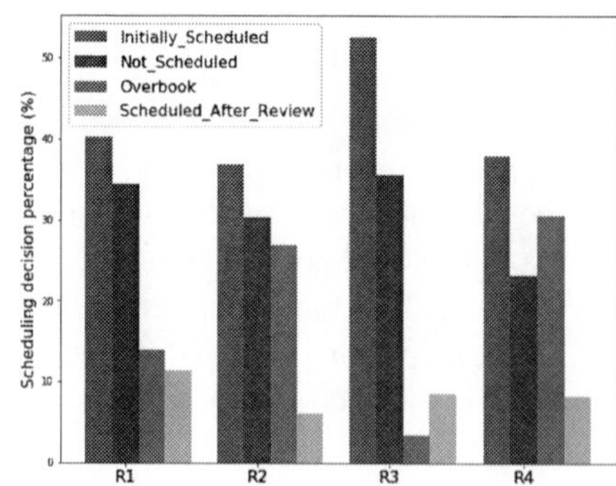

Figure 3: Distribution of scheduling decisions for different reviewers.

Question Type	D/M	P-C	P-EGD	P-Other
Linear SVM+BoN	0.71	0.75	0.81	0.32
Linear SVM+BoC	0.69	0.77	0.83	**0.34**
Kernel SVM+BoC	0.69	0.73	0.85	0.20
BERT	0.71	0.77	0.86	0.32
SciBERT	0.77	0.80	0.84	0.29
Bio+Clinical BERT	**0.78**	0.79	0.85	0.26
Bio+DS BERT	0.77	**0.84**	**0.92**	0.33

Table 3: F1 scores for question type classification with separate classifiers.

Question Type	D/M	P-C	P-EGD	P-Other
BERT	0.71	0.79	0.85	0.21
SciBERT	**0.82**	**0.86**	**0.89**	**0.41**
Bio+Clinical BERT	0.74	0.80	0.85	0.39
Bio+DS BERT	0.77	0.79	0.86	0.39

Table 4: F1 scores for question type classification with multi-task learning with different BERT variants.

FlexSig because there were only two instances in the split we evaluate on (current work is creating more annotations). The best overall performance was obtained by the SciBERT multi-task learning setup. In the single-task setting, Bio+Discharge Summary BERT alone provides several points of benefit on *Procedure-Colonoscopy* and *Procedure-EGD*. Multi-task learning provides an inconsistent benefit, increasing score in some categories while decreasing in others. However, when these two are combined, multi-task learning and SciBERT provide a large benefit over all other configurations.

5 Discussion & Conclusion

Within-reviewer results (diagonal of Table 2) indicate that predicting scheduling decisions from text alone is difficult, and there are few obvious cues to the urgency of a question. However, we also saw a large decrease in performance across reviewers, suggesting that individual reviewers behave very differently. Improving reviewer consistency may be a viable method for improving efficiency of specialist referrals in health systems. It still is not totally clear from these results whether the individual reviewers are inconsistent – it is possible that the classifier model we chose is simply the right representation to perform this task. Future work should look deeper at within-reviewer classifier performance to explore the degree to which scheduling decisions are essentially random.

One possible explanation for the improved performance of SciBERT is that it uses domain-specific pre-training as well as a domain-learned vocabulary (ClinicalBERT, in comparison, is pre-trained on clinical data but uses the original BERT vocabulary). Practically speaking, the result is that the SciBERT vocabulary contains more biomedical terms. For example, the term *colonoscopy* occurs as a single token in the SciBERT vocabulary, while the standard BERT vocabulary breaks it into several word pieces. We suspect that this makes it easier for SciBERT to learn domain-specific language, as the meaning is attached directly to the word piece embedding rather than being learned through BERT encoding layers.

Future work should explore further modeling of domain structure, including understanding question text better, but also in modeling relationships between output variables. For example, sometimes the Diagnosis/Management category is clear from expressions like *Please eval*, but in other cases the request is only implicit. In these cases, the best clue is the lack of any specific procedure request. A sequential classification decision process may be able to incorporate this logic. In addition, we are continuing the annotation process, including continuing to revise guidelines to improve agreement, annotating more questions for question type in the gastroenterology specialty, and developing guidelines for additional specialties. Our early results suggest that the question type classifier can still be improved with additional data, despite already-promising performance.

Acknowledgments

Research reported in this publication was supported by the National Institute On Minority Health And Health Disparities of the National Institutes of Health under Award Number R21MD012693. The content is solely the responsibility of the authors and does not necessarily represent the official views of the National Institutes of Health.

References

Emily Alsentzer, John Murphy, William Boag, Wei-Hung Weng, Di Jindi, Tristan Naumann, and Matthew McDermott. 2019. Publicly available clinical BERT embeddings. In *Proceedings of the 2nd Clinical Natural Language Processing Workshop*, pages 72–78, Minneapolis, Minnesota, USA. Association for Computational Linguistics.

Michael L. Barnett, Hal F. Yee, Ateev Mehrotra, and Paul Giboney. 2017. Los angeles safety-net program econsult system was rapidly adopted and decreased wait times to see specialists. *Health Affairs*, 36(3):492–499. PMID: 28264951.

Iz Beltagy, Kyle Lo, and Arman Cohan. 2019. Scibert: Pretrained language model for scientific text. In *EMNLP*.

Jacob Devlin, Ming-Wei Chang, Kenton Lee, and Kristina Toutanova. 2019. BERT: Pre-training of deep bidirectional transformers for language understanding. In *Proceedings of the 2019 Conference of the North American Chapter of the Association for Computational Linguistics: Human Language Technologies, Volume 1 (Long and Short Papers)*, pages 4171–4186, Minneapolis, Minnesota. Association for Computational Linguistics.

Alistair EW Johnson, Tom J Pollard, Lu Shen, H Lehman Li-wei, Mengling Feng, Mohammad Ghassemi, Benjamin Moody, Peter Szolovits, Leo Anthony Celi, and Roger G Mark. 2016. Mimic-iii, a freely accessible critical care database. *Scientific data*, 3:160035.

Jinhyuk Lee, Wonjin Yoon, Sungdong Kim, Donghyeon Kim, Sunkyu Kim, Chan Ho So, and Jaewoo Kang. 2020. Biobert: a pre-trained biomedical language representation model for biomedical text mining. *Bioinformatics*, 36(4):1234–1240.

Tomas Mikolov, Ilya Sutskever, Kai Chen, Greg S Corrado, and Jeff Dean. 2013. Distributed representations of words and phrases and their compositionality. In *Advances in neural information processing systems*, pages 3111–3119.

F. Pedregosa, G. Varoquaux, A. Gramfort, V. Michel, B. Thirion, O. Grisel, M. Blondel, P. Prettenhofer, R. Weiss, V. Dubourg, J. Vanderplas, A. Passos, D. Cournapeau, M. Brucher, M. Perrot, and E. Duchesnay. 2011. Scikit-learn: Machine learning in Python. *Journal of Machine Learning Research*, 12:2825–2830.

Jeffrey Pennington, Richard Socher, and Christopher Manning. 2014. Glove: Global vectors for word representation. In *Proceedings of the 2014 Conference on Empirical Methods in Natural Language Processing (EMNLP)*, pages 1532–1543, Doha, Qatar. Association for Computational Linguistics.

Matthew Peters, Mark Neumann, Mohit Iyyer, Matt Gardner, Christopher Clark, Kenton Lee, and Luke

Zettlemoyer. 2018. Deep contextualized word representations. In *Proceedings of the 2018 Conference of the North American Chapter of the Association for Computational Linguistics: Human Language Technologies, Volume 1 (Long Papers)*, pages 2227–2237, New Orleans, Louisiana. Association for Computational Linguistics.

Guergana K Savova, James J Masanz, Philip V Ogren, Jiaping Zheng, Sunghwan Sohn, Karin C Kipper-Schuler, and Christopher G Chute. 2010. Mayo clinical text analysis and knowledge extraction system (ctakes): architecture, component evaluation and applications. *Journal of the American Medical Informatics Association*, 17(5):507–513.

Thomas Wolf, Lysandre Debut, Victor Sanh, Julien Chaumond, Clement Delangue, Anthony Moi, Pierric Cistac, Tim Rault, R'emi Louf, Morgan Funtowicz, and Jamie Brew. 2019. Huggingface's transformers: State-of-the-art natural language processing. *ArXiv*, abs/1910.03771.

Towards Understanding ASR Error Correction for Medical Conversations

Anirudh Mani
Abridge AI Inc.
amani@abridge.com

Shruti Palaskar
Carnegie Mellon University
spalaska@cs.cmu.edu

Sandeep Konam
Abridge AI Inc.
san@abridge.com

Abstract

Domain Adaptation for Automatic Speech Recognition (ASR) error correction via machine translation is a useful technique for improving out-of-domain outputs of pre-trained ASR systems to obtain optimal results for specific in-domain tasks. We use this technique on our dataset of Doctor-Patient conversations using two off-the-shelf ASR systems: Google ASR (commercial) and the ASPIRE model (open-source). We train a Sequence-to-Sequence Machine Translation model and evaluate it on seven specific UMLS Semantic types, including Pharmacological Substance, Sign or Symptom, and Diagnostic Procedure to name a few. Lastly, we breakdown, analyze and discuss the 7% overall improvement in word error rate in view of each Semantic type.

1 Introduction

Off-the-shelf ASR systems like Google ASR are becoming increasingly popular each day due to their ease of use, accessibility, scalability and most importantly, effectiveness. Trained on large datasets spanning different domains, these services enable accurate speech-to-text capabilities to companies and academics who might not have the option of training and maintaining a sophisticated state-of-the-art in-house ASR system. However, for all the benefits these cloud-based systems provide, there is an evident need for improving their performance when used on in-domain data such as medical conversations. Approaching ASR Error Correction as a Machine Translation task has proven to be useful for domain adaptation and resulted in improvements in word error rate and BLEU score when evaluated on Google ASR output (Mani et al., 2020).

However, it is important to analyze and understand how domain adapted speech may vary from

Model	Transcript
Reference	you also have a *pacemaker* because you had sick sinus syndrome and it's under control
Google ASR	you also have a **taste maker** because you had sick sinus syndrome and it's under control
S2S	you also have a **pacemaker** because you had sick sinus syndrome and it's under control
Reference	like a heart disease uh *atrial fibrillation*
Google ASR	like a heart disease **asian populations**
S2S	like a heart disease **atrial fibrillation**

Table 1: Examples from Reference, Google ASR transcription and corresponding S2S model output for two medical words, *"pacemaker"* and *"atrial fibrillation"*. In this work, we investigate how adapting transcription to domain and context can help reduce such errors, especially with respect to medical words categorized under different Semantic types of the UMLS ontology.

ASR outputs. We approach this problem by using two different types of metrics - 1) overall transcription quality, and 2) domain specific medical information. For the first one, we use standard speech metric like word error rate for two different ASR system outputs, namely, Google Cloud Speech API[1] (commercial), and ASPIRE model (open-source) (Peddinti et al., 2015). For the second type of evaluation, we use the UMLS[2] ontology (O., 2004) and analyze the S2S model output for a subset of semantic types in the ontology using

---

[1] https://cloud.google.com/speech-to-text/
[2] The Unified Medical Language System is a collection of medical thesauri maintained by the US National Library of Medicine

Proceedings of the 1st Workshop on NLP for Medical Conversations, pages 7–11
Online, 10, 2020. ©2020 Association for Computational Linguistics
https://doi.org/10.18653/v1/P17

a variety of performance metrics to build an understanding of effect of the Sequence to Sequence transformation.

2   Related Work

While the need for ASR correction has become more and more prevalent in recent years with the successes of large-scale ASR systems, machine translation and domain adaptation for error correction are still relatively unexplored. In this paper, we build upon the work done by Mani et al. (Mani et al., 2020). However, D'Haro and Banchs (D'Haro and Banchs, 2016) first explored the use of machine translation to improve automatic transcription and they applied it to robot commands dataset and human-human recordings of tourism queries dataset. ASR error correction has also been performed based on ontology-based learning in (Anantaram et al., 2018). They investigate the use of including accent of speaker and environmental conditions on the output of pre-trained ASR systems. Their proposed approach centers around bio-inspired artificial development for ASR error correction. (Shivakumar et al., 2019) explore the use of noisy-clean phrase context modeling to improve ASR errors. They try to correct unrecoverable errors due to system pruning from acoustic, language and pronunciation models to restore longer contexts by modeling ASR as a phrase-based noisy transformation channel. Domain adaptation with off-the-shelf ASR has been tried for pure speech recognition tasks in high and low resource scenarios with various training strategies (Swietojanski and Renals, 2014, 2015; Meng et al., 2017; Sun et al., 2017; Shinohara, 2016; Dalmia et al., 2018) but the goal of these models was to build better ASR systems that are robust to domain change. Domain adaptation for ASR transcription can help improve the performance of domain-specific downstream tasks such as medication regimen extraction (Selvaraj and Konam, 2019).

3   Domain Adaptation for Error Correction

Using the reference texts and pre-trained ASR hypothesis, we have access to parallel data that is in-domain (reference text) and out-of-domain (hypothesis from ASR), both of which are transcriptions of the same speech signal. With this parallel data, we now frame the adaptation task as a translation problem.

Sequence-to-Sequence Models : Sequence-to-sequence (S2S) models (Sutskever et al., 2014) have been applied to various sequence learning tasks including speech recognition and machine translation. Attention mechanism (Bahdanau et al., 2014) is used to align the input with the output sequences in these models. The encoder is a deep stacked Long Short-Term Memory Network and the decoder is a shallower uni-directional Gated Recurrent Unit acting as a language model for decoding the input sequence into either the transcription (ASR) or the translation (MT). Attention-based S2S models do not require alignment information between the source and target data, hence useful for monotonic and non-monotonic sequence-mapping tasks. In our work, we are mapping ASR output to reference hence it is a monotonic mapping task where we use this model.

4   Experimental Setup

4.1   Dataset

We use a dataset of 3807 de-identified Doctor-Patient conversations containing 288,475 utterances split randomly into 230,781 training utterances and 28,847 for validation and test each. The total vocabulary for the machine translation task is 12,934 words in the ASR output generated using Google API and ground truth files annotated by humans in the training set. We only train word-based translation models in this study to match ASR transcriptions and ground truth with further downstream evaluations. To choose domain-specific medical words, we use a pre-defined ontology by Unified Medical Language System (UMLS) (O., 2004), giving us an exhaustive list of over 20,000 medications. We access UMLS ontology through the Quickumls package (Soldaini and Goharian, 2016), and use seven semantic types - Pharmacological Substance (PS), Sign or Symptom (SS), Diagnostic Procedure (DP), Body Part, Organ, or Organ Component (BPOOC), Disease or Syndrome (DS), Laboratory or Test Result (LTR), and Organ or Tissue Function (OTF). These are thereby referred by their acronyms in this paper. These seven semantic types were chosen to cover a spread of varied number of utterances available for each type's presence, from lowest (OTF) to the highest (PS) in our dataset.

Alignment: Since the ground truth is at utterance level, and ASR system output transcripts are

Ontology	Utts	Unique Terms
	Train, Test	Train, Test
PS	35301, 4481	1233, 532
DS	17390, 2191	859, 310
BPOOC	15312, 1944	513, 222
SS	14245, 1805	429, 181
DP	4016, 484	217, 82
LTR	3466, 407	70, 33
OTF	1866, 228	68, 26

Table 2: Breakdown of the Full Data based on REF.

Transcript	WER ($\Downarrow$)	BLEU ($\Uparrow$)
Google ASR output	41.0	52.1
+ S2S Adapted	34.1	56.4
ASPIRE ASR output	35.8	54.3
+ S2S Adapted	34.5	55.8

Table 3: Results for adaptive training experiments with Google ASR and ASPIRE model. We compare absolute gains in WER and BLEU scores with un-adapted ASR output.

at word level, specific alignment handling techniques are required to match the output of multiple ASR systems. This is achieved using utterance level timing information i.e., start and end time of an utterance, and obtaining the corresponding words in the ASR system output transcript based on word-level timing information (start and end time of each word). To make sure same utterance ID is used across all ASR outputs and the ground truth, we first process our primary ASR output transcripts from Google Cloud Speech API based on the ground truth and create random training, validation and test splits. For each ground truth utterance in these dataset splits, we also generate corresponding utterances from ASPIRE output transcripts similar to the process mentioned above. This results in two datasets corresponding to Google Cloud Speech and ASPIRE ASR models, where utterance IDs are conserved across datasets. However, this does lead to ASPIRE dataset having a lesser utterances as we process Google ASR outputs first in an effort maximize the size of our primary ASR model dataset.

Pre-trained ASR: We use the Google Cloud Speech API for Google ASR transcription and the JHU ASPIRE model (Peddinti et al., 2015) as two off-the-shelf ASR systems in this work. Google Speech API is a commercial service that charges users per minute of speech transcribed, while the ASPIRE model is an open-source ASR model. We explore the trends we observe in both–a commercial API as well as an open-source model.

5 Results and Discussions

5.1 Transcription Quality

We use WER and BLEU scores to evaluate improvement on ASR model outputs using the S2S model. A consistent gain is observed across all metrics, with an absolute improvement of 7% in WER and a 4 point absolute improvement in BLEU scores on Google ASR. While the Google ASR output can be stripped of punctuation for a better comparison, it is an extra post-processing step and breaks the direct output modeling pipeline. If necessary, ASPIRE model output and the references can be inserted with punctuation as well.

5.2 Qualitative Analysis

In Table 4, we compare S2S adapted outputs with Google ASR for each semantic type, broken down by Precision, Recall and F1 scores. The two outputs are also compared directly by counting utterances where S2S model made the utterance better with respect to a semantic term - it was present in the reference and S2S output but not Google ASR, and cases where S2S model made the utterance worse - semantic term was present in the reference and Google ASR but not S2S output. We refer to this metric as *semantic intersection* in this work.

As observed, the F1 scores are higher for S2S outputs for all the semantic types in the Ontology, except for one (BPOOC) where it ties. In terms of Precision and Recall too, S2S performs better for most categories. These numbers can be discussed with a couple of underlying factors - how common or rare the semantic terms are on average for each semantic type, and how many training examples has the model seen for those terms. This is important to consider as Google ASR learns on a much larger vocabulary of words spanning many different domains, where as S2S is trained on a domain specific dataset. For example, we see a large gain on Precision for DP, which can be attributed to the rarity of the terms under this category, like 'echocardiogram', 'pacemaker', etc. Its also for this reason we see only a slight improvement in Precision for PS even though it has the most number of training examples. Many of the medication names are rare, but a lot of them are pretty common

Ontology	Unique Terms	S2S adpt, ASR o/p			
	S, G, R	P	R	F1	SI
PS	282, 393, 532	**0.86** , 0.85	**0.61** , 0.55	**0.72** , 0.67	**0.10**, 0.02
DS	210, 302, 310	0.75 , 0.75	0.68 , 0.68	**0.76** , 0.75	**0.03**, 0.02
BPOOC	173, 235, 222	**0.82** , 0.81	0.70 , 0.70	0.75 , 0.75	0.02, 0.02
SS	144, 169, 181	0.87 , **0.88**	**0.74** , 0.72	**0.8** , 0.79	**0.03**, 0.01
DP	54, 73, 82	**0.89** , 0.75	0.65 , **0.70**	**0.75** , 0.72	0.02, **0.07**
LTR	26, 26, 33	0.77 , **0.85**	**0.67** , 0.61	**0.72** , 0.71	**0.07**, 0.01
OTF	26, 32, 26	**0.79** , 0.74	**0.79** , 0.77	**0.79** , 0.75	**0.04**, 0.02

Table 4: Medical WER results per Ontology for adaptive training experiments on Test data. We use Precision, Recall, F1 and Semantic Intersection (as defined in 5.2) metrics for comparing S2S model output to Google ASR.

nowadays even though they are domain specific, like 'aspirin'. Moreover, this is also supported by the numbers observed for BPOOC, where terms like 'legs', 'heart' and 'lungs' are the top 3 most frequently occurring words.

The number of unique terms for the S2S output are lower in comparison to Google ASR and reference as observed in Table 4. This might indicate that the S2S model is incorrectly modifying some Google ASR output medical terms which may not have as many examples in the Training set. However, our *semantic intersection* metric indicates that we get an overall improvement in all categories, except for DP. We hypothesize this to be largely due to a combination of how rare the words are, and the overall number of training examples for DP being low. When we calculate *semantic intersection* on the Full set, we get almost equal results for S2S and Google ASR outputs, 0.5 and 0.6 respectively. When we look at our top 5 and bottom 5 least frequent terms for each semantic types, almost all the terms overlap between S2S, Google ASR and reference, even though the number of unique terms might be less for S2S. Overall, it is evident from analyzing the results that as the number of occurrences increases for each medical term, the performance of the S2S model in identifying errors and correcting them increases rapidly, as shown in Table 2 and Table 4.

In a production environment, the S2S model may be confidently used for correcting ASR errors for top K most frequently occurring medical terms, where the value of K must be decided based on the dataset available for training. Future extension of this work will also be looking into the class imbalance problem for a more robust performance on different semantic types.

6 Conclusion

We present an analysis of how ASR Error Correction using Machine Translation impacts the different semantic types of the UMLS ontology for a medical conversation. We run the S2S model on a dataset of Doctor-Patient conversations as a post-processing step to optimize the Google off-the-shelf ASR system. We use different input representations and compare the performance of our S2S model using WER and BLEU scores on Google ASR and ASPIRE outputs. We deep dive into how our adaptation model affect medical WER for each semantic type, and breakdown the results using Precision, Recall, F1 and Semantic Intersection numbers between S2S and Google ASR. We establish the robustness of S2S model performance for more frequently occurring medical terms. In the future, we want to explore other representations like phonemes which might capture ASR errors better, and address the class imabalance problem for rarer medical terms in different semantic types.

Acknowledgments

We thank the University of Pittsburgh Medical Center (UPMC) and Abridge AI Inc. for providing access to de-identified data of Doctor-Patient conversations used in this work.

References

C Anantaram, Amit Sangroya, Mrinal Rawat, and Aishwarya Chhabra. 2018. Repairing asr output by artificial development and ontology based learning. In *IJCAI*, pages 5799–5801.

Dzmitry Bahdanau, Kyunghyun Cho, and Yoshua Bengio. 2014. Neural machine translation by jointly learning to align and translate. *CoRR*, abs/1409.0473.

Siddharth Dalmia, Xinjian Li, Florian Metze, and Alan W Black. 2018. Domain robust feature extraction for rapid low resource asr development. In *2018 IEEE Spoken Language Technology Workshop (SLT)*, pages 258–265. IEEE.

Luis Fernando D'Haro and Rafael E Banchs. 2016. Automatic correction of asr outputs by using machine translation.

Anirudh Mani, Shruti Palaskar, Nimshi Venkat Meripo, Sandeep Konam, and Florian Metze. 2020. Asr error correction and domain adaptation using machine translation. In *2020 IEEE International Conference on Acoustics, Speech and Signal Processing (ICASSP)*. IEEE.

Zhong Meng, Zhuo Chen, Vadim Mazalov, Jinyu Li, and Yifan Gong. 2017. Unsupervised adaptation with domain separation networks for robust speech recognition. *arXiv preprint arXiv:1711.08010*.

Bodenreider O. 2004. The unified medical language system (umls): integrating biomedical terminology. Nucleic Acids Res. 2004 Jan 1;32(Database issue):D267-70. doi: 10.1093/nar/gkh061. PubMed PMID: 14681409; PubMed Central PMCID: PMC308795. Nucleic Acids Res.

Vijayaditya Peddinti, Guoguo Chen, Vimal Manohar, Tom Ko, Daniel Povey, and Sanjeev Khudanpur. 2015. Jhu aspire system: Robust lvcsr with tdnns, ivector adaptation and rnn-lms. In *2015 IEEE Workshop on Automatic Speech Recognition and Understanding (ASRU)*, pages 539–546. IEEE.

Sai P Selvaraj and Sandeep Konam. 2019. Medication regimen extraction from medical conversations. *arXiv preprint arXiv:1912.04961*.

Yusuke Shinohara. 2016. Adversarial multi-task learning of deep neural networks for robust speech recognition. *Proc. Interspeech 2016*.

Prashanth Gurunath Shivakumar, Haoqi Li, Kevin Knight, and Panayiotis Georgiou. 2019. Learning from past mistakes: improving automatic speech recognition output via noisy-clean phrase context modeling. *APSIPA Transactions on Signal and Information Processing*, 8.

Luca Soldaini and Nazli Goharian. 2016. Quickumls: a fast, unsupervised approach for medical concept extraction. In *MedIR workshop, sigir*, pages 1–4.

Sining Sun, Binbin Zhang, Lei Xie, and Yanning Zhang. 2017. An unsupervised deep domain adaptation approach for robust speech recognition. *Neurocomputing*, 257:79–87.

Ilya Sutskever, Oriol Vinyals, and Quoc V Le. 2014. Sequence to sequence learning with neural networks. In *Advances in Neural Information Processing Systems 27*, pages 3104–3112. Curran Associates, Inc.

Pawel Swietojanski and Steve Renals. 2014. Learning hidden unit contributions for unsupervised speaker adaptation of neural network acoustic models. In *Spoken Language Technology Workshop (SLT), 2014 IEEE*, pages 171–176. IEEE.

Pawel Swietojanski and Steve Renals. 2015. Differentiable pooling for unsupervised speaker adaptation. In *Acoustics, Speech and Signal Processing (ICASSP), 2015 IEEE International Conference on*, pages 4305–4309. IEEE.

Studying Challenges in Medical Conversation with Structured Annotation

Nan Wang
Hunan University
nwang.christine@gmail.com

Yan Song
Sinovation Ventures
clksong@gmail.com

Fei Xia
University of Washington
fxia@uw.edu

Abstract

Medical conversation is a central part of medical care. Yet, the current state and quality of medical conversation is far from perfect. Therefore, a substantial amount of research has been done to obtain a better understanding of medical conversation and to address its practical challenges and dilemmas. In line with this stream of research, we have developed a multi-layer structure annotation scheme to analyze medical conversation, and are using the scheme to construct a corpus of naturally occurring medical conversation in Chinese pediatric primary care setting. Some of the preliminary findings are reported regarding *1) how a medical conversation starts, 2) where communication problems tend to occur, and 3) how physicians close a conversation.* Challenges and opportunities for research on medical conversation with NLP techniques will be discussed.

1 Introduction

Medical conversation is at the core of medical care. Through conversation, doctors collect the information needed to form a diagnosis and provide a treatment recommendation for the patient's condition. Effective communication is essential for achieving optimal medical outcomes. Yet breakdowns in doctor-patient conversation are common in medical practices. For example, the largest proportion of hospital and community health services complaints in the UK were about communications with medical professional in 2017-2018 (NHS, 2018). Thus, a better understanding of medical conversation (e.g., how it is conducted; what practical problems and dilemmas doctors and patients face) could improve not only the quality of care, but also the efficiency of the healthcare system.

In this paper, we first review the major issues that medical conversation research has investigated; we then introduce the data and methods that we use to analyze medical conversation; lastly, we present some preliminary findings based on our analysis of the corpus and conclude with a discussion on implications of the study and our future work.

2 Research on Medical Conversation

Research on medical conversation has a long tradition. Earlier studies that use naturally occurring medical conversation data can be traced back to 1970s in the United Kingdom and the United States. Having audio-recorded and analyzed 2500 recordings of medical conversation, British researchers Byrne and Long (1976) were regarded as the pioneers in medical conversation research. At about the same time, American physicians Korsch and Negrete (1981) conducted one of the most influential studies on medical communication, based on 800 audio-recordings of conversation collected in the Los Angeles Children's Hospital. These studies showed that medical communication practices significantly affect patient health outcomes.

More recently, a substantial body of research using conversation analysis (Drew et al., 2001; Heritage and Maynard, 2006) emerged and investigated a wide range of topics in medical conversation. These topics generally fall into three categories.

How is medical conversation conducted? This stream of research examines the *process* and *constituent activities* of medical conversation. In other words, how do physician and patient coordinate in this social encounter and how is medical conversation organized?

Unlike many other types of conversation, medical conversation are treated as having a discernable overall structure. For example, acute primary care encounters in the UK were found ordinarily beginning with an opening sequence, progressing through problem presentation, history taking,

Proceedings of the 1st Workshop on NLP for Medical Conversations, pages 12–21
Online, 10, 2020. ©2020 Association for Computational Linguistics
https://doi.org/10.18653/v1/P17

physical examination, diagnosis, and treatment recommendations, and then onto a closing sequence (Byrne and Long, 1976).

While this overall structure of conversation can be considered as socialized through physicians' training in medical school and patients' repeated exposure since childhood, it is also considered a product of coordination and negotiation between physicians and patients in local interaction context. Therefore, a lot of research in this stream has been done to investigate questions such as how one activity transits toward another in medical conversation, and what constitutes a complete sequence within some particular activity phase.

For example, studies find that there are varying expectations for what constitutes a complete diagnosis sequence and a treatment recommendation sequence. While patients' no response or weak response (*e.g., mm hmm*) is treated as sufficient for a diagnosis sequence to complete (Heritage and Sefi, 1992), patients' explicit acceptance of treatment recommendation is oriented as necessary for completing a treatment recommendation sequence (Stivers, 2005).

What are the practical problems and dilemmas in medical conversation? The second stream of research on medical conversation concerns with more concrete problems in medical conversation. For instance, how physicians' and patients' *actions* are designed and *sequences* are organized to deal with various kinds of practical challenges and dilemmas in medical conversation.

For example, physicians use various forms of questions to solicit patients' problem presentations, and these different action types can afford different opportunities for patients' contribution to medical conversation. Specifically, the length of patients' problem presentation is significantly longer if physicians use open-ended solicitation questions, as compared with close-ended questions (Heritage and Robinson, 2006).

In addition, when delivering a diagnosis to terminal patients, it is found that physicians can deploy a rather complicated form of sequence (the so-called *News Delivery Sequence (NDS)*), in which patients' perspectives of their health condition are incorporated, in order to prepare them for the bad news (Maynard, 1997).

How does medical conversation affect medical outcomes? The third stream of medical conversation research examines the associations between doctor-patient interaction and outcomes in medical conversation, such as patient satisfaction or medication adherence.

For instance, research shows that if a candidate diagnosis (e.g., *I don't know whether she's got a sinus, but she has a lot of drainage in her nose.*) appears in patients' problem presentation, physicians are more likely to perceive the patient as expecting antibiotics (Stivers et al., 2003), and are more likely to prescribe antibiotics inappropriately (Mangione-Smith et al., 1999).

Similarly, toward the end of medical conversation, it is found that physicians' action design has a significant impact on the effect of medical communication. For instance, by replacing the word *any* in '*Do you have **any** other concerns you want to address today?*' with the word *some*, it reduces the likelihoods of patients' unaddressed concerns by up to 50% (Heritage and Robinson, 2006).

In sum, a substantial amount of effort has been made to obtain a better understanding of medical conversation. As a result, many important issues have been discovered and effective solutions have been provided to improve the practice of medical communication and the quality of care.

However, the existing resources and tools for medical conversation research suffer from two major problems, which put significant obstacles to the advancement of the field. These problems are: 1) collecting and analyzing natural conversation data in the medical setting requires a tremendous amount of resources (e.g., labor, time); 2) few standard coding frameworks exist, which allow for systematic analysis of medical conversation that takes into account the interactivity of utterances. Although coding schemes such as Roter Interaction Analysis System (RIAS) (Roter and Larson, 2002) attempt to implement an exhaustive classification of the events in medical conversation, these schemes tend to treat utterances in conversation as isolated units (Heritage and Maynard, 2006).

Motivated by these considerations, we construct a corpus that consists of 1,000 medical conversations and develop a coding scheme that captures the deep structure of conversation. Then we conduct a systematic microanalysis of the medical conversation, which takes into account of both the content and the context of utterances. While not being a focus of this paper, we will conclude the paper with a few potential use cases of our proposed work.

3 Data and Corpus

In this section, we introduce the the corpus that we construct for medical conversation research.

3.1 Video-recorded Data

A total of 1,000 medical conversation were video-recorded in Chinese pediatric primary care settings. Participants involve 14 physicians and 1,000 patients with their caregivers in 9 hospitals in northern, central and eastern China.

For each conversation, a complete course of medical consultation is included, starting from the patients getting seated, progressing through the discussion of patients' health conditions, and toward the patients leaving the office.

Topics in the conversation are mostly concerning children's acute respiratory tract infections (ARTIs) problems, which involve health complaints such as fever, cough, etc.

The conversation is primarily between physicians and caregivers, similar to pediatric primary care conversation in other countries. It should be noted that although most of the conversation is dyadic, multi-party conversation is also common in our corpus, as more than one caregiver can be present and contribute to conversation.

3.2 Transcribed Data

The video-recorded medical conversation data are transcribed manually by trained research assistants. Adopting the Conversation Analysis transcribing conventions (Jefferson, 2004), each conversation is segmented into turns at turn-taking positions. Besides capturing the verbatim of each turn, the transcription also captures a series of para-linguistic features (e.g., dysfluencies, intonations, overlaps of turns, noticeable silence in and between turns, non-verbal actions such as nodding, etc.), which are essential aspects of natural spoken language.

In addition, the transcribed text is automatically segmented into words using an in-house CRF word segmenter trained on the Chinese Treebank (Xia et al., 2000), so as to provide the necessary basis for conducting related NLP tasks.

3.3 Ethical Considerations

All research procedures were reviewed and approved by UCLA IRB and UW IRB. All identifying information (e.g., person, institution, location names) has been removed from the corpus.

3.4 Our Analysis

In this paper, we describe a series of findings based on the corpus regarding the following aspects.

Overall organization and opening a medical conversation Similar to ordinary conversation, medical conversation is a social encounter where physicians and patients build rapport and social relationship (Schegloff, 1968). Thus, quite often at the beginning of medical conversation, physicians and patients engage in social exchange activities such as greetings and identifying. We refer to this kind of exchange as *opening phase* in medical conversation. Ex 1 illustrates an example of *opening phase* in our corpus.[1]

Ex 1: Opening in conversation
1 **DOC:** Hi. @NAME@? How are you?
2 **DAD:** Yes, that's us. How are you?
3 **DOC:** What's going on today?

However, medical conversation is also where patients and physicians deal with patients' health concerns. Thus, in many cases, conversation starts with physicians and patients talking about the patients' health problems, without going through *opening phase*. We refer to this kind of activity as *problem presentation phase*.

Although medical conversation can be opened with either the *opening phase* or the *problem presentation phase*, there seems to be a distributional difference in different cultural and medical settings. We will discuss this in more detail in Section 5.

Sequence expansion and making treatment decisions A second aspect of our analysis focuses on *sequences* within some particular phases in medical conversation.

For example, within *treatment recommendation phase*, we examined how treatment recommendations are delivered and received. Specifically, it is found that treatment decisions (e.g., antibiotic prescriptions) can be negotiated between physicians and patients by patients withholding acceptance of physicians' recommendations. Thus, the minimal form of 'recommendation-acceptance' sequence can be expanded quite extensively, in order to secure the patients' explicit acceptance of the physicians' treatment recommendation. Ex 2 and 3 illustrate examples of a non-expanded form and

---

[1]To save space, we omit the Chinese line and show the translation only

an expanded form of treatment recommendation sequence, respectively.

Ex 2: Non-expanded treatment sequence
1 **DOC:** I'll probably put her on some antibiotics.
2 **MOM:** Okay.

Ex 3: Expanded treatment sequence
1 **DOC:** I'll give her some decongestant.
2 **MOM:** Decongestant?
3 **DOC:** Yeah. It's a viral infection. Antibiotics won't kill.
4 **MOM:** Okay, let's use some decongestant.

Although arguably the most ideal form of treatment recommendation sequence consists of two turns (like in Ex 2), patients' acceptance may be delayed and will then be pursued by physicians (like in Ex 3).

Action design and initiating conversation closures Similar to ordinary conversation, the physician and the patient also need to coordinate to close a medical conversation, rather than simply falling silent (Sacks and Schegloff, 1973).

In medical conversation, past research has shown that, upon reaching a point when the treatment decision is made, physicians produce various forms of actions to initiate the closure of the medical visit (West, 2006) . Although the patients can always resist such attempts to close and the conversation may go back-and-forth to other phases, the conversation is considered as entering the *closing phase*, when these closure initiation actions are produced.

After physicians secure the warrant from patients to terminate the conversation, the two parties can then proceed to the terminal exchange of the conversation. Ex 4 illustrates an example of the closure initiation action in our corpus.

Ex 4: Closure (a) Making future arrangement
1 **DOC:** Okay. Follow up in two days, ok?
2 **DAD:** Okay. Thank you.
3 **DOC:** You're welcome.

In Ex 4, the physician initiates the closure of the medical conversation by making a future arrangement for the patient's follow-up visit at line 1. The patient's father accepts the proposal and the two parties immediately proceed to the terminal exchange (*thank you-you're welcome*) at lines 2-3.

Besides the closure initiation action *(a) making*

future arrangement shown in Ex 4, there are two other forms of action recurrently observed in the medical conversation: *(b) summarizing treatment plans* and *(c) announcing closures*. Ex 5 and 6 illustrate two examples.

Ex 5: Closure (b) Summarizing treatment plans
1 **DOC:** Just use these three medications, ok?
2 **MOM:** Okay
3 **DOC:** Alright.
4 **MOM:** Ok. Goodbye, doctor.

Ex 6: Closure (c) Announcing closures
1 **DOC:** Okay. That's it.
2 **MOM:** Thank you, Doctor.
3 **DOC:** You're welcome.

The above examples show that, when implementing particular actions in conversation, there can be different turn designs so as to accommodate the particular contingencies arising from the interaction context. These choices of turn design may afford different opportunity for the recipient's participation, and thereby have different impact on the subsequent development of the conversation. A close examination of this phenomenon thus provides a window to uncover the practical constraints that the patients and the physicians face. Actionable solutions can then be developed to deal with these constraints.

4   COSTA Scheme

To enable a systematic analysis of medical conversation in the dimensions that we described above, we developed an annotation scheme that marks up the structure of medical conversation at multiple levels. In addition, application-dependent labels can be created and added on top of the structural annotation, tailored to particular researchers' interests.

Below we briefly introduce how medical conversations are analyzed using the COnversational STructures and Actions (COSTA) scheme, in terms of 1) conversational structures, and 2) application-dependent labels for conversational actions.

4.1   Annotating conversation structure

Figure 1 illustrates how the hierarchical structure of medical conversation is annotated according to the COSTA scheme. Detail of the COSTA scheme can be found in (Wang et al., 2018).

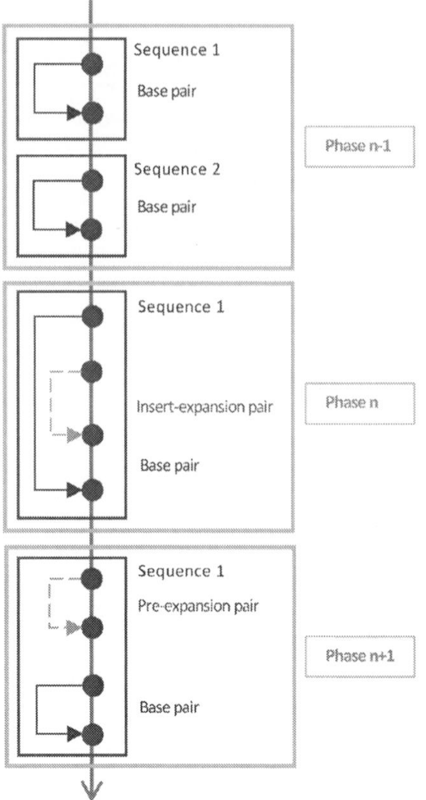

Figure 1: A hierarchical organization of medical conversation. The blue dots represent turns lined up in the temporal order of a conversation. The yellow boxes represent *phases* in the medical conversation (e.g., opening, history-taking, treatment recommendation, etc.), which are consisted of one or more sequences. The blue boxes represent *sequences*, which can be minimally consists of one pair, or multiple pairs with one base pair and its expansion pairs. Within a sequence, the red arrows link the two turns of a base *pair*; whereas the gray arrows suggest that the two connected turns belong to an expansion pair, which is dependent on the base pair.

Overall organization At the highest level, medical conversation is segmented and marked up with a series of component *phases*. Based on related findings on the overall organization of medical conversation from past research (Robinson, 2003; Byrne and Long, 1976) and analysis of our corpus, labels of the phases are created, including, *(P0) opening, (P1) problem presentation, (P2) history taking, (P3) physical examination, (P4) diagnosis, (P5) treatment recommendation, (P6) lifeworld discussion, (P7) closing, (P8) additional health concerns.* Disruptions (e.g., physicians interrupted by calls) are also common in medical conversation and we mark up them as *(P9) unrelated activities.* [2]

Despite that past research has shown these constituent phases are ordered in a normative sense,

it is not unusual that the physicians and patients go back and forth between these phases. By annotating which phase a turn belongs to, we are not only able to show where the boundaries are among the phases in medical conversation, but also how transitions are coordinated by the participants.

Pair and turn dependency Unlike many other types of discourse, conversation is interactive in nature. Thus, turns in conversation cannot be understood alone. Instead, each turn should be understood regarding whether they set up an expectation for a next turn or fulfill the expectation set up by a prior turn. Pairs of turns which are linked by conditional relevance is referred to as *adjacency pairs* and considered the building block of conversation (Sacks et al., 1974; Schegloff, 2007).

Based on this idea, turns are annotated with respect to which turn they are connected to, and within a pair, a first pair part is distinguished from its second part (e.g., question vs. answer, request vs. grant, greeting vs. return greeting, etc.).

This type of dependency relationship between two turns in conversation has been attended to in related work such as the SWBD-DAMSL coding scheme (*forward-communicative-function and backward-communication-function*) (Jurafsky et al., 1997; Core and Allen, 1997). It has demonstrated significant value for NLP tasks such as dialog act modeling (Stolcke et al., 2000).

Sequence and pair dependency As mentioned above, *adjacency pair* is considered the most basic unit of conversation (Sacks et al., 1974; Schegloff, 2007) . It is also considered as the minimal form of *sequence* in conversation. In the most ideal scenario, a sequence is complete with one base pair (e.g., a question gets its answer, an invitation gets its acceptance), as shown in Ex 2. But more commonly, pairs are expanded to accommodate various types of contingencies in interaction (e.g., repairing a problem of hearing, checking understanding) (Schegloff, 1980, 1992, 1997) , as shown in Ex 3.

In these cases, several pairs cluster into a coherent sequence, with one base pair and multiple expansion pairs dependent on it. Based on the sequential position of the base pair and the expansion pairs, there are pre-expansions, insert-expansions, and post-expansions (Schegloff, 2007) .

Based on this idea, the COSTA scheme marks up the dependency relationship between pairs and distinguishes the base pairs from the expansion

---

[2]Labels of phases can be adjusted for different types of medical conversation.

pairs. In Figure 1, we illustrate a sequence with an insert-expansion pair in *phase n-1*, and a sequence with a pre-expansion pair in *phase n+1* .

In sum, the hierarchical annotation scheme of the COSTA describes the deep structure of medical conversation. It thus allows for systematic study of conversation at multi-level granularity, including *phases*, *sequences*, *pairs*, and *turns*.

4.2 Application-dependent labels

According to researchers' specific research interests and application scenarios, additional labels can be created and added to any particular level of conversation (e.g., *phase, sequence, pair, turn)*. Systematic analyses of the labeled data in the medical conversation corpus can help provide answers to various kind of research questions.

In this study, we ask the following questions: (1) How does a medical conversation start? (2) Where do communication problems tend to occur? (3) How do physicians close a conversation?

To answer question (1), we examine the labels of the initial *phase* of the medical conversation. If a conversation is opened with physicians and patient caregivers identifying and greeting each other, it is annotated as *(P0) Opening phase*; if the opening involves physician asking and/or patient presenting health problems, it is marked as *(P1) Problem presentation*.

To answer question (2), we examine the organization of *sequence*. Sequences that consist of only one base pair without any expansions are considered as produced with less difficulty. This is compared with sequences that consist of multiple pairs, with the base pair expanded with several dependent pairs. Communication problems tend to occur in phases where there are more expanded sequences.

To answer question (3), we create labels to distinguish different types of physicians' closure initiation actions. Based on past research on closing in medical conversation (West, 2006) and preliminary analysis of our data, physicians' closure initiation actions can be classified into three types: *(a) making future arrangement, (b) summarizing treatment plans*, and *(c) announcing closures*. Examples of the three types of action are in Ex 4-6.

Item	Number
Number of Visits	187
Number of Hospitals	5
Number of Physicians	9
Number of Patients	187
Average length of a visit	4.9 minutes

Table 1: Meta information of the subset in this study.

Item	Total	Average per visit
Characters	275,303	1472.2
Words	158,798	849.2
Turns	23,060	123.3
Pairs	11,833	63.3
Sequences	5,359	28.7

Table 2: Statistics of the subset in this study. Total number of visits in this subset is 187.

5 Results

5.1 Corpus Statistics

As it is an ongoing project, here we present some statistics based on a subset of acute visits in the corpus. Table 1 shows the meta data of the subset. Table 2 shows the statistics of the transcribed data of this subset. All the experimental results in this section are based on this subset.

5.2 How does medical conversation start?

Since medical conversation is both a social encounter where relationship is built and a task-oriented activity organized with a clear goal, we find that medical conversation in our corpus start with either the *(P0) Opening phase* or the *(P1) Problem presentation phase*. Table 3 describes the distribution of the two types of conversation opening in our dataset. As shown in the table, a majority of the conversation starts with participants going directly to discuss the health problem of patients.

This is compared with ordinary conversation, in which the initial exchanges almost always involve a *summons-answer* sequence (SA sequence) (Schegloff, 1968). Typical SA sequences include *telephone ring–hello, Johnny?–Yes, Bill–looks up*, etc. After the channel for communication is established through the SA sequence, the conversationalists then proceed to the reason for the talk.

When comparing with the findings on conversation opening in the American primary care, we find that there is a small variance in the distribution of the two types of opening. In the American primary care conversation, it is reported that less than 10% of the cases are opened with the *(P0) Opening phase* (Heritage and Robinson, 2006; Robinson

17

Conversational opening type	Count	%
(P0) Opening phase	62	33
(P1) Problem presentation phase	125	67
Total	187	100

Table 3: Distribution of the two types of conversational opening in the subset.

Phase type	Seq # per phase	Turn # per seq
(P0) Opening phase	1.67	2.15
(P1) Problem presentation phase	3.11	3.95
(P2) History-taking phase	6.50	4.71
(P3) Physical examination phase	3.03	3.21
(P4) Diagnosis phase	2.12	4.49
(P5) Treatment phase	5.32	6.63
(P6) Lifeworld discussion phase	3.18	3.01
(P7) Closing phase	2.14	3.72
(P8) Additional problem phase	0.83	4.13

Table 4: Average number of sequences in each phase of medical conversation and average number of turns in each sequence in those phases. The total number of conversation is 187. *(P9) Unrelated activities phase* is not included in this table, as they do not directly contribute to the understanding or progressivity of the conversation.

and Heritage, 2005). However, in the Chinese medical conversation, 33% of the cases are opened with *(P0)* (see Table 3). Thus, in a greater proportion of the Chinese medical conversation, the participants do engage in social activities, such as identity confirmation, greetings, or even 'intimacy ploy'.

It should be noted that although the findings from the Chinese pediatric primary care are not directly comparable to that in the American primary care setting, this distributional variance in conversational opening highlights the difference in the norms and service procedures of medical interaction in two cultures. Specifically, while patients in the American primary care are normally received by nurses or medical assistants first in their medical visits, patients are directly seen by their physicians in the Chinese consultation room. The higher proportion of the *opening phase* in the Chinese corpus thus can be explained by the practical constraints that physicians have to confirm patients' identity at the beginning of the medical consultation.

5.3 Where do problems tend to occur?

When examining the *process and overall organization* of the medical conversation, we find that there are considerable variances in the shape of various *phases*. Table 4 shows the average number of sequences and turns in each type of phase in the subset of the corpus.

Among all the phases, the *treatment recommen-*

dation phase is where sequences are most likely to be expanded. Specifically, a sequence in the *treatment recommendation phase* takes an average of 6.63 turns to complete. In comparison, the average number of turns for a sequence in the *problem presentation* take 3.95 turns; and that number is the lowest in the *opening phase*, averaging 2.15 turns.

Looking into the *sequences* in the treatment recommendation phase, it is observed that physicians' treatment recommendations are not always immediately accepted by patient caregivers in the next turn. In face of such patient resistance, physicians must to pursue caregivers' acceptance, and the sequence continues to expand until the patients' explicit acceptance is displayed (as shown in Ex 3).

In addition, in our prior work, we labeled and analyzed the caregivers' *actions* that they use to overtly advocate for antibiotic treatment in the *treatment recommendation phase*. The results showed that, when caregivers use one or more of the following actions *a) explicit requests for antibiotics, b) statements of desire for antibiotics, c) inquiries about antibiotics, and d) evaluations of treatment effectiveness,* the likelihoods of them receiving antibiotic prescriptions from the physicians increased by over 9 times (Odds Ratio = 9.23, 95% Confidence Interval = 3.30-33.08) (Wang et al., 2018). This finding corroborates the fact that antibiotic over-prescription is prevalent in the Chinese pediatric primary care (Li et al., 2012), and parental pressure on physicians in medical conversation plays a significant role in antibiotic over-prescription (Stivers et al., 2003; Mangione-Smith et al., 1999).

5.4 How do physicians close a conversation?

Closing a medical conversation is a delicate matter, as physicians and patients may have conflicting agendas. While patients may still have unmentioned concerns, physicians may have to terminate the conversation so as to move to the next patient.

To deal with such practical challenge, we find that physicians use several types of actions to initiate the closure of medical conversation. These actions include: *a) making arrangement for future activities, b) summarizing the topic-in-progress,* and *c) announcing closures.* Table 5 illustrates the relative distribution of these three types of action design in our corpus. Examples of the three types of actions are shown in Ex 4-6 in Section 3.

Compared with closing in ordinary conversation,

Closure initiation actions	Count	%
(a) Making future arrangements	78	52
(b) Summarizing treatment plans	57	38
(c) Announcing closures	15	10
Total	150	100

Table 5: Closure initiation actions and their distributions in the Chinese medical conversation. The total number of cases in this table is 150. In the remaining 37 cases, closures are initiated by caregivers and are excluded from the analysis.

the range of actions that physicians use to initiate medical conversation closures are highly similar. After the topical closure attempts are accepted by the caregiver or the patient, participants move on to the pre-closing sequence, in which they pass the floor to one another and confirm there is nothing more to talk about. Once the warrant to terminate is established, they move on to the termination sequence, in which they exchange farewell (bye–bye), display appreciation (thanks–you're welcome), or acknowledge the closure of the conversation (ok–ok) (Sacks and Schegloff, 1973).

When comparing our findings with related findings in the American primary care, we find that there exist some important variances. Besides the three types of actions that the Chinese physicians use, there is another type of action observed in the US data: *checking patients' unmet concerns (e.g., 'Do you have some other problems that you want to talk about?')* (West, 2006).

Again, although the findings from the Chinese pediatric primary care are not directly comparable to that in the American primary care, this difference in the range of action designs that the physicians use highlights the practical problems and constraints that exist in the Chinese pediatric setting. In the Chinese medical setting, and urban tertiary hospitals in particular, physicians are commonly overloaded (Hu and Zhang, 2016). In a day, a physician could see as many as 100 patients, and the length of the medical conversation tend to be very short, averaging 4.9 minutes for each conversation in our corpus. Absence of this action (i.e., checking patients' unmet concerns) in the Chinese corpus can be partially attributable to this.

6 Discussion

In this section, we discuss several potential use cases of this study.

6.1 Facilitating conversational understanding

One reason that conversation understanding is difficult is because the meaning of utterances often depends on the context. For instance, the word *yeah* as a response to a yes-no question is doing the action of agreeing. In contrast, the word *yeah* uttered by a speaker when another speaker is in the middle of a long stretch of talk may indicate that the former is listening to the latter; it does not mean that the former agrees with the latter.

Moreover, if we treat the conversation simply as a sequence of turns without internal structure, multi-turn understanding may not be easily achieved. The idea of internal structure is that turns in conversation are not like beads on a string; instead, they are organized in coherent clusters. As a result, the two turns within an adjacency pair are not always adjacent. For instance, in Ex 3, Lines 1 and 4 form an adjacency pair, with the word *Okay* in Line 4 responding to the treatment recommendation made in Line 1. Lines 2-3 in between form another adjacency pair, which is an insert-expansion pair of the base pair. These dependent pairs form up one coherent sequence, and sequences of similar kind form up a coherent phase in conversation.

Thus, the whole conversation is represented as a tree structure, similar to dependency structure for a sentence. Compared with treating the conversation as a sequence of turns, having such tree structure information would make it much easier to infer that the word *Okay* in Line 4 indicates the acceptance of the recommendation in Line 1. In this sense, the conversational structural information helps multi-turn conversational understanding.

Manually annotating such a tree structure is labor intensive and time consuming, but once such a corpus is created, automatic tools can be trained on the corpus, the same way that dependency parsers are trained on treebanks. The tools can then be used to process new conversations. Our corpus thus is the first conversation treebank annotated with conversational structures and actions according to the COSTA scheme. In the scheme, the label set (e.g., phase labels and action labels) is application-dependent, whereas the structure levels (e.g., *phase, sequence, pair, turn*) should remain mostly the same for many applications.

6.2 Extracting information from medical conversation

Due to the nature of medical conversation, there are often natural correspondences between phases in medical conversation and sections in Electronic Health Record (EHR). For example, *problem pre-*

sentation phase in the medical conversation corresponds to *symptom section* in the EHR; *treatment phase* in the medical conversation corresponds to *prescription section* in the EHR, etc. Therefore, the medical conversation structures and labels (e.g., phase types) could provide valuable cues when building NLP systems for tasks such as information extraction. For instance, prescribed medication is more likely to appear in the *treatment phase* of medical conversation, rather than *opening* or *closing phase*. While the *history-taking phase* may also contain medication names, such medication concerns primarily with the medication history of the patient, rather than the medication prescribed during the current visit.

Apart from information extraction, the structural representation of medical conversation can help other NLP tasks such as automatic summarization of patient medical visit. Building high-quality BioNLP systems for such tasks has great potential to reduce physicians' workload and increase the time they spend on treating patients.

6.3 Conducting more communication-related research with automatically processed data

Our current study looks at some of the major issues in medical communication. For future work, we plan to apply the same methodology to other issues in medical communication and conversations in other domains.

While the current study relies on manual annotation of the conversation, once NLP tools have been trained on annotated data (as described above), we can use the tools to analyze a large amount of new conversations automatically, and significantly speed up the analytical process of conversation.

7 Conclusion

In this paper, we introduced some of the major issues that existing medical conversation research has focused on; we described the data that we use for conducting medical conversation research. To analyze medical conversation more systematically, we proposed an annotation scheme, which can capture the hierarchical structure of the medical conversation and be extended to include application-specific labels. Based on a subset of the annotated data, we report findings regarding how medical conversation is opened and closed in the Chinese pediatric consultations and how one can identify

places that problems tend to occur.

This study makes several contributions to medical conversation research. First, to our best knowledge, the corpus that we are constructing is the first medical conversation dataset with structural annotation. It is a valuable resource for conducting medical communication research, and can also be used to train NLP systems such as a conversation parser. Second, COSTA is a general scheme for annotating conversational structures and actions. The annotation facilitates systematic analysis of medical conversation and there are other potential use cases as outlined in the previous section. While we use the scheme to build a Chinese corpus consisting of medical conversations, COSTA can be applied to conversation in other domains or in other languages.

For future work, we will finish annotations of the corpus and release it to the public. We will start training NLP tools in order to evaluate the usefulness of the corpus for the use cases mentioned above.

References

Patrick S. Byrne and Berrie E. Long. 1976. *Doctors Talking to Patients: A Study of the Verbal Behaviour of General Practitioners Consulting in Their Surgeries*. H.M. Stationery Office, London, UK.

Mark Core and James Allen. 1997. Coding Dialogs with the DAMSL Annotation Scheme. In *Proceedings of AAAI Fall Symposium on Communicative Action in Humans and Machines*.

Paul Drew, John Chatwin, and Sarah Collins. 2001. Conversation analysis: A method for research into interactions between patients and health-care professional. *Health Expectations*, 4(1):58–70.

John Heritage and Douglas W. Maynard. 2006. *Communication in Medical Care: Interaction between Primary Care Physicians and Patients*. Cambridge University Press, Cambridge.

John Heritage and Jeffrey D. Robinson. 2006. The structure of patients' presenting concerns: Physicians' opening questions. *Health Communication*, 19(2):89–102.

John Heritage and Sue Sefi. 1992. Dilemmas of advice: Aspects of the delivery and reception of advice in interactions between health visitors and first time mothers. In Paul Drew and John Heritage, editors, *Talk at Work*, pages 3–65. Cambridge University Press, Cambridge.

Yinhuan Hu and Zixia Zhang. 2016. Skilled doctors in tertiary hospitals are already overworked in china. *The Lancet Global Health*, 3(12):e737.

Gail Jefferson. 2004. Glossary of transcript symbols with an introduction. In Gene H. Lerner, editor, *Conversation Analysis: Studies from the First Generation*, chapter 2, pages 13–31. John Benjamins, Amsterdam / Philadelphia.

Daniel Jurafsky, Elizabeth Shriberg, and Debra Biasca. 1997. Switchboard SWBD-DAMSL Shallow-Discourse-Function Annotation Coders Manual, Draft 13. Technical report, University of Colorado, Boulder.

Barbara M. Korsch and Vida F. Negrete. 1981. Doctor-patient communication. *Scientific American*, 227(2):66–74.

Yongbin Li, Jing Xu, Fang Wang, Bin Wang, Liqun Liu, Wanli Hou, Hong Fan, Yeqing Tong, Juan Zhang, and Zuxun Lu. 2012. Overprescribing in china, driven by financial incentives, results in very high use of antibiotics, injections, and corticosteroids. *Health Affairs (Project Hope)*, 31(5):1075–1082.

Rita Mangione-Smith, Elizabeth McGlynn, Marc N. Elliott, Paul Krogstad, and Robert H. Brook. 1999. The relationship between perceived parental expectations and pediatrician antimicrobial prescribing behavior. *Pediatrics*, 103(4):711–718.

Douglas W. Maynard. 1997. The news delivery sequence: Bad news and good news in conversational interaction. *Research on Language and Social Interaction*, 30(2):93–130.

NHS. 2018. Data on Written Complaints in the NHS 2017-18. Technical report, Health and Social Care Information Center.

Jeffery D. Robinson. 2003. An interactional structure of medical activities during acute visits and its implications for patient's participation. *Health Communication*, 15(1):27–57.

Jeffrey D. Robinson and John Heritage. 2005. The structure of patients' presenting concerns: the completion relevance of current symptoms. *Journal of the Association for Computing Machinery*, 61(2):481–493.

Debora Roter and Susan Larson. 2002. The roter interaction analysis system (rias): Utility and flexibility for analysis of medical interactions. *Patient Education and Counseling*, 42:243–251.

Harvey Sacks and Emanuel A. Schegloff. 1973. Opening up closings. *Semiotica*, 8(4):289–327.

Harvey Sacks, Emanuel A. Schegloff, and Gail Jefferson. 1974. A Simplest Systematics for the Organization of Turn-taking for Conversation. *Language*, 50(4, Part 1):696–735.

Emanuel A. Schegloff. 1968. Sequencing in Conversational Openings. *American Anthropologist*, 70(6):1075–1095.

Emanuel A. Schegloff. 1980. Preliminaries to Preliminaries: "Can I Ask You a Question?". *Sociological Inquiry*, 50(3-4):104–152.

Emanuel A. Schegloff. 1992. Repair After Next Turn: The Last Structurally Provided Defense of Intersubjectivity in Conversation. *American Journal of Sociology*, 97(5):1295–1345.

Emanuel A. Schegloff. 1997. Third Turn Repair. In R. Gregory Guy, Crawford Feagin, Deborah Schiffrin, and John Baugh, editors, *Towards a Social Science of Language: Papers in Honor of William Labov. Volume 2: Social Interaction and Discourse Structures*, page 31–40. John Benjamins.

Emanuel A. Schegloff. 2007. *Sequence Organization in Interaction: Volume 1: A Primer in Conversation Analysis*. Cambridge University Press.

Tanya Stivers. 2005. Parent resistance to physicians' treatment recommendations: One resource for initiating a negotiation of the treatment decision. *Health Communication*, 18(1):41–74.

Tanya Stivers, Rita Mangione-Smith, Marc N. Elliott, Laurie McDonald, and John Heritage. 2003. Why do physicians think parents expect antibiotics? what parents report vs what physicians believe. *The Journal of Family Practice*, 52(2):140–148.

Andreas Stolcke, Klaus Ries, Noah Coccaro, Elizabeth Shriberg, Rebecca Bates, Daniel Jurafsky, Paul Taylor, Rachel Martin, Carol Van Ess-Dykema, and Marie Meteer. 2000. Dialogue Act Modeling for Automatic Tagging and Recognition of Conversational Speech. *Computational Linguistics*, 26(3):339–373.

Nan Wang, Yan Song, and Fei Xia. 2018. Coding structures and actions with the COSTA scheme in medical conversations. In *Proceedings of the BioNLP 2018 workshop*, pages 76–86, Melbourne, Australia. Association for Computational Linguistics.

Candace West. 2006. Coordinating Closing in Primary Care Visits: Producing Continuity of Care. In John Heritage and Douglas W. Maynard, editors, *Communication in Medical Care: Interaction between Primary Care Physicians and Patients*, pages 379–414. Cambridge University Press.

Fei Xia, Martha Palmer, Nianwen Xue, Mary Ellen Okurowski, John Kovarik, Fudong Chiou, Shizhe Huang, Tony Kroch, and Mitch Marcus. 2000. Developing Guidelines and Ensuring Consistency for Chinese Text Annotation. In *Proceedings of the Second Language Resources and Evaluation Conference (LREC)*.

Generating Medical Reports from Patient-Doctor Conversations using Sequence-to-Sequence Models

**Seppo Enarvi[1], Marilisa Amoia[1], Miguel Del-Agua Teba[1], Brian Delaney[1],
Frank Diehl[1], Guido Gallopyn[1], Stefan Hahn[1], Kristina Harris[1], Liam McGrath[1],
Yue Pan[1], Joel Pinto[1], Luca Rubini[1], Miguel Ruiz[1], Gagandeep Singh[1],
Fabian Stemmer[1], Weiyi Sun[1], Paul Vozila[1], Thomas Lin[2], Ranjani Ramamurthy[2]**

[1]Nuance Communications, 1 Wayside Road, Burlington, MA 01803
`firstname.lastname@nuance.com`
[2]Microsoft Corporation, One Microsoft Way, Redmond, WA 98052
`{tlin,ranjanir}@microsoft.com`

Abstract

We discuss automatic creation of medical reports from ASR-generated patient-doctor conversational transcripts using an end-to-end neural summarization approach. We explore both recurrent neural network (RNN) and Transformer-based sequence-to-sequence architectures for summarizing medical conversations. We have incorporated enhancements to these architectures, such as the pointer-generator network that facilitates copying parts of the conversations to the reports, and a hierarchical RNN encoder that makes RNN training three times faster with long inputs. A comparison of the relative improvements from the different model architectures over an oracle extractive baseline is provided on a dataset of 800k orthopedic encounters. Consistent with observations in literature for machine translation and related tasks, we find the Transformer models outperform RNN in accuracy, while taking less than half the time to train. Significantly large wins over a strong oracle baseline indicate that sequence-to-sequence modeling is a promising approach for automatic generation of medical reports, in the presence of data at scale.

1 Introduction

There has been an increase in medical documentation requirements over the years owing to increased regulatory requirements, compliance for insurance reimbursement, caution over litigation risk, and more recently towards increased patient participation. According to a study on 57 U.S. physicians, for every hour with a patient, a physician takes an additional hour of personal time doing clerical work (Sinsky et al., 2016). Increased documentation burden has been identified as one of the main contributing factors for physician burnout (Wright and Katz, 2018). In another, larger study, U.S. physicians who used electronic health records (EHRs) or computerized physician order entry (CPOE) were found to be less satisfied with the time spent on administrative work (Shanafelt et al., 2016).

Increased physician burnout not only affects the health and well-being of the physicians, it can also lead to increased medical errors, increased job turnover, reduced productivity, and reduced quality of patient care (Panagioti et al., 2017). Factors related to physician burnout and its consequences have been studied in detail in the literature (Patel et al., 2018b).

Use of automatic speech recognition (ASR) to dictate medical documentation has contributed significantly to the efficiency of physicians in creating narrative reports (Payne et al., 2018). However the content of the report has already been discussed with the patient during the encounter. Medication list and orders entered into the EHRs are also discussed with the patient. In other words, creation of medical documentation by the physician may be viewed as a redundant task given that the content is already discussed with the patient.

There has been a surge in research on automatic creation of medical documentation from patient-doctor conversations. A lot of it is focused on extracting medical information and facts from the patient-doctor conversation (Happe et al., 2003; Quiroz et al., 2019). This could involve extracting clinical standard codes (Leroy et al., 2018), clinical entities such as symptoms, medications, and their properties (Du et al., 2019a,b), or medical regimen (Selvaraj and Konam, 2020). This extracted information could then be used to generate a report (Finley et al., 2018a,b). Such information extraction systems require creating an annotated conversation corpus (Patel et al., 2018a; Shafran et al., 2020).

For example, the NLP pipeline described by Finley et al. (2018a) first extracts knowledge from an

Proceedings of the 1st Workshop on NLP for Medical Conversations, pages 22–30
Online, 10, 2020. ©2020 Association for Computational Linguistics
https://doi.org/10.18653/v1/P17

ASR transcript and then generates the report. The knowledge extraction consists of tagging speaker turns and sentences with certain classes using RNN-based models, and using techniques such as string matching, regular expressions, and data-driven supervised and unsupervised approaches to extract information from the tagged sentences. This is followed by data-driven templates and finite state grammars for report generation.

We take a different approach where the problem is cast as translation (source language is conversational in nature and target language is clinical) and summarization (input contains redundant and irrelevant information, and target is a concise and precise note) at the same time. Given recent advances in neural transduction technology (Bahdanau et al., 2015; See et al., 2017; Vaswani et al., 2017), we explore the end-to-end paradigm for generating medical reports from ASR transcripts. This eliminates the need for annotated corpora that are required for training intermediate processing steps. As a result this approach is scalable across various medical specialties.

Sequence-to-sequence models have been used for summarizing radiology notes into the short *Impressions* section, possibly incorporating also other domain-specific information (Zhang et al., 2018; MacAvaney et al., 2019). In contrast, our system creates a report directly from the conversation transcript. Disadvantages of the end-to-end approach include that it limits the ability to inject prior knowledge and audit system output, and may potentially result in inferior performance.

2 Dataset

We use data consisting of ambulatory orthopedic surgery encounters. Speaker-diarized conversation transcripts corresponding to the audio files were obtained using an automatic speech recognizer. The reports for orthopedic surgery are organized under four sections—history of present illness (HPI), physical examination (PE), assessment and plan (AP), and diagnostic imaging results (RES). The HPI section captures the reason for visit, and the relevant clinical and social history. The PE section captures both normal and abnormal findings from a physical examination. The RES section outlines impressions from diagnostics images such as X-ray and CT scans. Finally, the AP section captures the assessment by the doctor and treatment plan e.g. medications, physical therapy etc.

	Size	Source		Target	
		Avg	Max	Avg	Max
Ortho HPI	802k	961	7,008	116	2,920
Ortho RES	444k	993	6,873	48	878
Ortho PE	769k	970	7,008	128	1,456
Ortho AP	811k	967	7,008	160	2,639
CNN&DM	287k	681	2,496	48	1,248
XSum	204k	431	33,161	23	432

Table 1: Statistics of our orthopedic report creation task and two other summarization tasks. Number of training examples and average and maximum number of tokens in the source and target sequence.

Experimental results are reported on a dataset that consists of around 800k encounters from 280 doctors. The dataset is partitioned chronologically (date of collection) into train, validation and evaluation partitions. The evaluation partition includes 4,000 encounters from 80 doctors. The doctors present in the evaluation set are present in the train set. Since the models do not require supervision outside the workflow, this paradigm is scalable, though future work will assess generalization to unseen doctors. We only use non-empty examples for training and evaluation. The RES section is empty in about 50 % of the examples.

Table 1 shows more detailed statistics of our dataset in terms of the number of training examples and source and target sequence lengths. The table also shows corresponding statistics for two prominent datasets for abstractive summarization that were not used in this study: CNN and Daily Mail, as processed by Nallapati et al. (2016), and XSum (Narayan et al., 2018). As shown in the table both the source and target sequences in our data are significantly longer than in the standard databases.

3 Modeling

We use neural sequence-to-sequence models for summarizing the doctor-patient conversations. The input to the model is a sequence of tokens generated by the speech recognizer. The medical reports consist of four sections, and we produce each section using a separate sequence-to-sequence model.

The task closely resembles machine translation, so we use models that are similar to neural machine translation models. There are, however, several differences from a typical machine translation task:

1. The source and target sequences are in the same language, thus we can use the same vo-

cabulary for input and output.

2. Report generation may require reasoning over a long span of sentences. The sequences, especially the source sequences, are significantly longer, since we cannot translate sentences separately.

3. Information may be incomplete (patient gesturing where it hurts), redundant (patient or doctor repeating information), or irrelevant conversation. In translation the semantic content in both the source and the target sequence is the same.

The models that we use are based on the encoder-decoder architecture that is well known from neural machine translation (Sutskever et al., 2014). All the models rely on attention (Bahdanau et al., 2015). The encoder creates context-dependent representations of the input tokens, and the decoder produces the next-token probability from those representations. During inference the output is generated autoregressively using the next-token probabilities.

Very long sequences increase the memory usage and training time, and make it more difficult to learn the model parameters. We truncate the source sequences to 2,000 tokens and the target sequences to 500 tokens during training. 10 % of the source sequences and 0.1 % of the target sequences were above this threshold. During inference we truncate the inputs to 3,000 tokens. Only 4 % of the test examples were originally longer than this limit.

In this work we compare models that are based on recurrent neural networks and models based on Transformer (Vaswani et al., 2017). The following sections describe these models and the enhancements that we have implemented.

3.1 RNN with Attention

The RNN sequence-to-sequence model with attention was introduced by Bahdanau et al. (2015). The encoder creates context-dependent input representations using a bidirectional RNN. The decoder produces the next-token probability using a unidirectional RNN, since future information is not available. We used LSTM (Hochreiter and Schmidhuber, 1997) as the recurrency mechanism.

We included in the model some of the enhancements from the RNMT+ model (Chen et al., 2018)—dropout, residual connections, layer normalization, and label smoothing. We also increased the number of encoder and decoder layers to two, but further increasing the number of layers did not give any benefit. We did not see significant benefit from using multi-head attention.

3.2 Hierarchical Encoder

Training the RNN model is slow due to their inherently sequential form precluding parallelization within the long input and output sequences. With longer input sequences it also becomes increasingly difficult for the model to learn to attend to relevant parts of the input.

Inspired by Cohan et al. (2018), we split the input sequence into 8 equal-length segments that are encoded independently. The segments can be processed in parallel, speeding up training considerably. The final LSTM hidden state from forward and backward directions of each segment are concatenated and projected into a segment embedding. After a stack of segment encoders, one more bidirectional LSTM runs over the segment embeddings of the previous layer (see Figure 1). The attention distribution is computed using both the token-level (second layer) and segment-level (third layer) outputs similar to Cohan et al. (2018)—the token-level scores are weighted by the normalized segment-level scores.

The hierarchical encoder sped up training by a factor of three with little to no impact on the summarization accuracy.

3.3 Pointer-Generator

To facilitate effective copying of parts of conversations to the output we implemented the pointer-generator network from See et al. (2017). It reuses the encoder-decoder attention distribution as a pointer for the copy mechanism. The same attention distribution is still used for computing the context vector and a probability distribution over the output vocabulary. The attention distribution is taken as a probability distribution over the input tokens and interpolated with the vocabulary distribution. The interpolation coefficient is learnt from the context vector, decoder LSTM state, and decoder input embedding. This mechanism also enables handling of words that are not present in the decoder vocabulary.

The pointer-generator network is illustrated in Figure 1. The context vector from attention is fed into a linear layer with the decoder state to produce the vocabulary distribution. The attention distribution is interpolated with the vocabulary distribution,

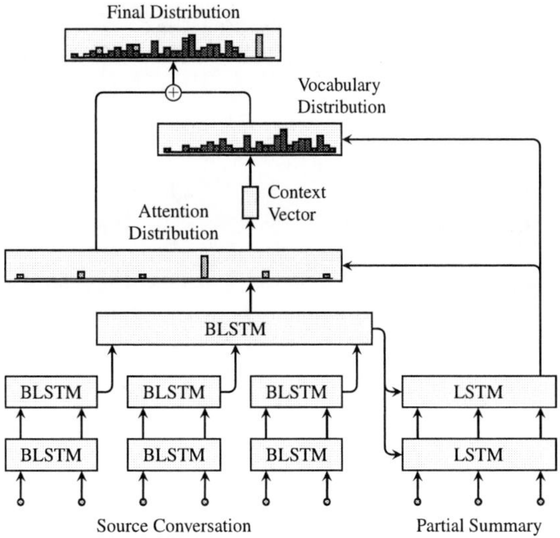

Figure 1: Illustration of the RNN sequence-to-sequence model with hierarchical encoder and pointer-generator copy mechanism. Segments of source conversation are encoded independently using two bidirectional LSTM layers, and a third layer runs over the final segment embeddings. The attention distribution is computed using both the token-level and segment-level outputs. The final distribution is interpolated from the attention distribution and the vocabulary distribution using a predicted coefficient.

although the connections for predicting the interpolation coefficient have been omitted from the figure for clarity.

The authors also introduce a coverage loss for training. They define coverage as the sum of attention weights over previous decoding steps. The coverage loss penalizes for attending to positions where the coverage is already high. The purpose is to encourage the model to attend to all input positions while decoding a sequence, and reduce repetition. We found the coverage loss to be somewhat helpful with a small weight (0.001).

3.4 Transformer

Transformer uses self-attention (Vaswani et al., 2017) in the encoder and decoder to create context-dependent representations of the inputs. In our experiments both encoder and decoder consist of six layers of self-attention. Each decoder layer attends to the top of the encoder stack after the self-attention. Additionally each encoder and decoder layer contains a position-wise feed-forward or convolutional network that consists of two transformations and a ReLU activation in between. The

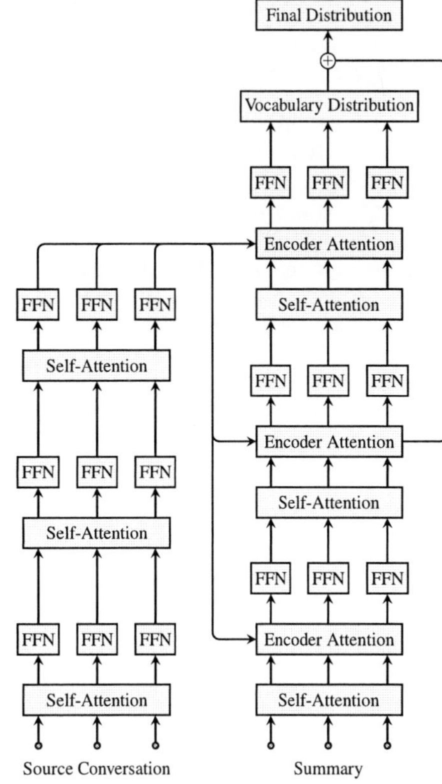

Figure 2: Illustration of the Transformer sequence-to-sequence model with pointer-generator copy mechanism. Each encoder layer consists of self-attention and a position-wise feed-forward network. Decoder layers also attend to the top of the encoder stack. We take one attention distribution from the penultimate decoder layer and interpolate it with the vocabulary distribution using a predicted coefficient. The final distribution includes the vocabulary tokens and the present source tokens. Layer normalization and residual connections are omitted for clarity.

fact that these layers can be computed in parallel for every position makes training more efficient than training RNN models.

Following Vaswani et al. (2017), we use the base model size, i.e. 8 attention heads with a total of 512 outputs and a 2048-dimensional feed-forward network. Following Domhan (2018), we apply layer normalization (Ba et al., 2016) before the self-attention and feed-forward sub-layers. This greatly stabilizes training and speeds up convergence with long inputs, confirming observations earlier made with deep networks (Wang et al., 2019).

Since the output of attention is independent of the order of the inputs, we inject position-dependent information into the inputs. In the origi-

	ROUGE-L RERR				Fact F_1 RERR			
	HPI	RES	PE	AP	HPI	RES	PE	AP
RNN	4.5	38.8	50.1	18.7	27.5	44.6	64.9	24.2
Hierarchical RNN	9.2	43.3	56.3	21.4	29.7	49.7	68.4	26.8
Hierarchical RNN + PG	9.2	45.4	53.7	22.3	30.8	51.3	67.1	29.2
Transformer	18.6	49.6	**65.4**	40.2	39.1	55.0	**74.6**	46.8
Transformer + PG	**19.2**	**51.0**	**65.4**	**42.0**	**39.5**	**56.7**	74.2	**49.4**

Table 2: Relative error rate reductions calculated from ROUGE-L and fact extractor F_1 scores. The scores are relative to an oracle baseline model that produces the longest common subsequence between the input and the reference output. The models labeled with PG use the pointer mechanism and coverage training loss.

nal paper, Vaswani et al. (2017) added sinusoidal position information before the first layer. We use relative position representations (Shaw et al., 2018) that are added inside the attention mechanism, which we found to work slightly better.

We also implemented a pointing mechanism in the Transformer model, similar to the RNN pointer-generator. For pointing we can use any distribution over the source tokens. The Transformer model creates several encoder-decoder attention distributions, one for each attention head in each layer. In principle any single head or the average of heads could be used for pointing. We argue that dedicating a single attention head should be sufficient, since the parameters of that head will be trained to attend to the tokens that are good candidates for copying. In this case the rest of the attention heads will not be affected and will perform their usual function, unlike when averaging over the attention heads. The penultimate layer seems to naturally learn alignments (Garg et al., 2019), so we use its first attention head for pointing. A simplified picture of the model is in Figure 2.

	HPI	RES	PE	AP
RNN	168	168	168	168
Hierarchical RNN	168	117	168	168
Hierarchical RNN + PG	168	131	168	168
Transformer	68	26	68	64
Transformer + PG	69	26	70	66

Table 3: Training time in hours on the four report sections for the various model architectures. Training time was restricted to one week, causing most RNN jobs to stop before reaching the maximum number of training steps.

4 Experiments

We train the models on Azure cloud using NVIDIA V100 GPUs. Each training job is distributed to 8 GPUs. We use data-parallel training, i.e. each GPU processes their share of the mini-batch and then the gradients are averaged over the GPU devices. The batch size is set to a maximum of 7,000 source tokens per GPU. NVIDIA NCCL library is used to perform the communication efficiently.

We use a vocabulary consisting of the 10k most frequent words. The same vocabulary is shared between the source and target tokens.

We use Nesterov's Accelerated Gradient (Nesterov, 1983) with the RNN models, while Adam (Kingma and Ba, 2015) is found to perform better with the Transformer models. We train the models a maximum of 400k steps, excepting RES section models, which are trained until 200k steps due to their fewer examples and shorter targets. This corresponds to approximately 25 epochs on RES section and 30 epochs on other sections. During this time we observe that training has practically converged and training longer would not provide significant benefit. We also limit individual model training to one week as a cost control.

Improved performance is obtained via averaging model parameter from 8 checkpoints, with interval length as a function of total training steps. Where helpful, we use a cyclical learning rate schedule, with the cycle length set to the checkpoint saving interval, so that the saved checkpoints would correspond to the minimums of the learning rate schedule (Izmailov et al., 2018).

4.1 Results

ROUGE (Lin, 2004) is a collection of metrics designed for evaluation of summaries. We calculate ROUGE-L, which is an F_1 score that is based on the lengths of the longest common subsequences

Partial ASR Transcript

[doctor]: we'll do a celebrex refill let me see you back four to six months earlier if needed okay hey good to see you good to see you [patient]: thank you thank you thank you

Reference Output

i have refilled her celebrex to have available. the patient will follow up in four to six months or earlier if needed.

Baseline Model Output

celebrex four to six months earlier if needed

Transformer PG Model Output

i have provided the patient with a refill of celebrex. the patient will follow up in 4-6 months or sooner if needed.

Figure 3: An excerpt of a speaker-diarized ASR transcript, its reference AP section output, the baseline model output, and partial output of the Transformer pointer-generator model (all lower-cased, without formatting). The baseline model produces the longest common subsequence between the transcript and the reference output.

between the reference and hypothesis sentences. We have noticed that it measures the fluency of the language well. However, we are also interested in assessing factual correctness. For this we utilize a proprietary machine-learning-based clinical fact extractor. It is capable of extracting medical facts such as conditions and medications, as well as their attributes such as body part, severity, or dosage. We extract facts from the model output and the ground-truth report, and compute the F_1 score from these two sets.

We publish our scores relative to an oracle baseline model, which extracts the longest common subsequence between the input conversation and the reference output. An example of such output is in Figure 3. Table 2 shows the relative error rate reduction (RERR) from ROUGE-L and fact extractor F_1 scores. We define the error rate as the complement $(1 - s)$ of the original score.

The Transformer models clearly obtain better scores than any of the RNN models. Partly this is because most RNN experiments are limited by the

	HPI	RES	PE	AP
RNN	12.1	0.4	3.8	19.9
Hierarchical RNN	2.0	0.2	1.2	15.8
Hierarchical RNN + PG	3.6	0.1	1.3	16.9
Transformer	2.0	0.1	1.2	0.3
Transformer + PG	1.7	1.0	0.8	0.3

Table 4: Percentage of model output considered part of a repetition. We define repetition as a sentence that occurs at least four times in the same report, or an n-gram of at least 16 tokens that repeats consecutively.

maximum training time. In the RES section both hierarchical RNN and Transformer models reached 200k training steps, but Transformer performance is still superior. An example output generated by the Transformer pointer-generator model is shown in Figure 3.

The training times are shown in Table 3. All but one of the RNN experiments were stopped after reaching the one week limit. Normal RNN training was terminated after approximately 50k steps, while hierarchical RNN progressed 160k–230k steps over the same duration. Performance was similar at an equal number of steps, but given fixed practical time and cost constraints, the hierarchical encoder yields improved results.

The pointer mechanism generally provided a small performance boost, with the largest improvements in RES and AP section quality. Interestingly, the pointer mechanism can even hurt performance in PE section. This is exaggerated with the RNN models by the fact that the pointer-generator model is slower to train and progressed only 175k steps, while the same model without the pointer mechanism and coverage loss progressed 225k steps in the time limit. Generally the ROUGE-L and fact F_1 scores seem correlated, displaying similar differences across models.

4.2 Repetition

By visual inspection of generated reports we noticed that some models suffer from an excessive amount of repetitions. We identified two main categories: sentences that occur multiple times in the same report and consecutively repeating n-grams. We try to assess the amount of repetition in model output by detecting these two types of patterns. Not all occurrences of such patterns are mistakes, however, and even the reference targets contain such patterns. We limit to sentences that occur at least 4 times and repeating n-grams that are at least 16 tokens long. Table 4 shows the repetition rates in model outputs as the percentage of tokens that fall into either of these categories. The table shows that the problem diminishes when training longer.

In the reference reports there are only a few instances of tokens that we consider repetitive, and these appear to be mistakes by the writer of the report. We should then aim at 0 % repetition rate. Note that the purpose of this metric is not to detect language where for example the same frequent words are used more often than in natural language. We rather wanted to assess how widely the models suffer from artificial and clearly erroneous repetition of word sequences.

5   Conclusions

In this paper we compared RNN and Transformer-based sequence-to-sequence architectures for medical report generation from patient-doctor conversations. This study demonstrates the ability of sequence-to-sequence models, in particular Transformer, to not only extract relevant clinical conversation excerpts, but abstractively summarize in a relatively fluent and factually correct medical report. Especially when working within compute and time budgets, Transformer is superior to traditional RNN-based models, and scalable to large datasets.

Visual inspection showed that commonly occurring problems in the generated reports included repeated sentences and hallucinated clinically consistent sentences unfounded by the conversations. Minimally a human would need to be in the loop to verify or correct these machine-generated reports. Future work includes comparing end-to-end approaches with a pipeline of clinical information extraction and natural language generation methods.

References

Lei Jimmy Ba, Jamie Ryan Kiros, and Geoffrey E. Hinton. 2016. Layer normalization. In *NIPS 2016 Deep Learning Symposium*, Barcelona, Spain.

Dzmitry Bahdanau, Kyunghyun Cho, and Yoshua Bengio. 2015. Neural machine translation by jointly learning to align and translate. In *3rd International Conference on Learning Representations, ICLR 2015, San Diego, CA, USA, May 7-9, 2015, Conference Track Proceedings*.

Mia Xu Chen, Orhan Firat, Ankur Bapna, Melvin Johnson, Wolfgang Macherey, George F. Foster, Llion Jones, Mike Schuster, Noam Shazeer, Niki Parmar,

Ashish Vaswani, Jakob Uszkoreit, Lukasz Kaiser, Zhifeng Chen, Yonghui Wu, and Macduff Hughes. 2018. The best of both worlds: Combining recent advances in neural machine translation. In *Proceedings of the 56th Annual Meeting of the Association for Computational Linguistics, ACL 2018, Melbourne, Australia, July 15-20, 2018, Volume 1: Long Papers*, pages 76–86. Association for Computational Linguistics.

Arman Cohan, Franck Dernoncourt, Doo Soon Kim, Trung Bui, Seokhwan Kim, Walter Chang, and Nazli Goharian. 2018. A discourse-aware attention model for abstractive summarization of long documents. In *Proceedings of the 2018 Conference of the North American Chapter of the Association for Computational Linguistics: Human Language Technologies, Volume 2 (Short Papers)*, pages 615–621, New Orleans, Louisiana. Association for Computational Linguistics.

Tobias Domhan. 2018. How much attention do you need? A granular analysis of neural machine translation architectures. In *Proceedings of the 56th Annual Meeting of the Association for Computational Linguistics (Volume 1: Long Papers)*, pages 1799–1808, Melbourne, Australia. Association for Computational Linguistics.

Nan Du, Kai Chen, Anjuli Kannan, Linh Tran, Yuhui Chen, and Izhak Shafran. 2019a. Extracting symptoms and their status from clinical conversations. In *Proceedings of the 57th Annual Meeting of the Association for Computational Linguistics*, pages 915–925, Florence, Italy. Association for Computational Linguistics.

Nan Du, Mingqiu Wang, Linh Tran, Gang Lee, and Izhak Shafran. 2019b. Learning to infer entities, properties and their relations from clinical conversations. In *Proceedings of the 2019 Conference on Empirical Methods in Natural Language Processing and the 9th International Joint Conference on Natural Language Processing (EMNLP-IJCNLP)*, pages 4979–4990, Hong Kong, China. Association for Computational Linguistics.

Gregory Finley, Erik Edwards, Amanda Robinson, Michael Brenndoerfer, Najmeh Sadoughi, James Fone, Nico Axtmann, Mark Miller, and David Suendermann-Oeft. 2018a. An automated medical scribe for documenting clinical encounters. In *Proceedings of the 2018 Conference of the North American Chapter of the Association for Computational Linguistics: Demonstrations*, pages 11–15, New Orleans, Louisiana. Association for Computational Linguistics.

Gregory Finley, Erik Edwards, Amanda Robinson, Najmeh Sadoughi, James Fone, Mark Miller, David Suendermann-Oeft, Michael Brenndoerfer, and Nico Axtmann. 2018b. An automated assistant for medical scribes. In *Proceedings of Interspeech 2018*, pages 3212–3213.

Sarthak Garg, Stephan Peitz, Udhyakumar Nallasamy, and Matthias Paulik. 2019. Jointly learning to align and translate with Transformer models. In *Proceedings of the 2019 Conference on Empirical Methods in Natural Language Processing and the 9th International Joint Conference on Natural Language Processing (EMNLP-IJCNLP)*, pages 4453–4462, Hong Kong, China. Association for Computational Linguistics.

Andr Happe, Bruno Pouliquen, Anita Burgun, Marc Cuggia, and Pierre [Le Beux]. 2003. Automatic concept extraction from spoken medical reports. *International Journal of Medical Informatics*, 70(2):255 – 263. MIE 2002 Special Issue.

Sepp Hochreiter and Jürgen Schmidhuber. 1997. Long short-term memory. *Neural computation*, 9(8):17351780.

Pavel Izmailov, Dmitrii Podoprikhin, Timur Garipov, Dmitry P. Vetrov, and Andrew Gordon Wilson. 2018. Averaging weights leads to wider optima and better generalization. In *Proceedings of the Thirty-Fourth Conference on Uncertainty in Artificial Intelligence, UAI 2018, Monterey, California, USA, August 6-10, 2018*, pages 876–885. AUAI Press.

Diederik P. Kingma and Jimmy Ba. 2015. Adam: A method for stochastic optimization. In *Proceedings of the 3rd International Conference on Learning Representations (ICLR)*.

Gondy Leroy, Yang Gu, Sydney Pettygrove, Maureen K Galindo, Ananyaa Arora, and Margaret Kurzius-Spencer. 2018. Automated extraction of diagnostic criteria from electronic health records for autism spectrum disorders: Development, evaluation, and application. *Journal of Medical Internet Research*, 20(11):e10497.

Chin-Yew Lin. 2004. ROUGE: A package for automatic evaluation of summaries. In *Text Summarization Branches Out*, pages 74–81, Barcelona, Spain. Association for Computational Linguistics.

Sean MacAvaney, Sajad Sotudeh, Arman Cohan, Nazli Goharian, Ish Talati, and Ross W. Filice. 2019. Ontology-aware clinical abstractive summarization. In *Proceedings of the 42nd International ACM SIGIR Conference on Research and Development in Information Retrieval*, SIGIR19, pages 1013–1016, New York, NY, USA. Association for Computing Machinery.

Ramesh Nallapati, Bowen Zhou, Cicero dos Santos, Çağlar Gulçehre, and Bing Xiang. 2016. Abstractive text summarization using sequence-to-sequence RNNs and beyond. In *Proceedings of The 20th SIGNLL Conference on Computational Natural Language Learning*, pages 280–290, Berlin, Germany. Association for Computational Linguistics.

Shashi Narayan, Shay B. Cohen, and Mirella Lapata. 2018. Don't give me the details, just the summary! Topic-aware convolutional neural networks for extreme summarization. In *Proceedings of the 2018 Conference on Empirical Methods in Natural Language Processing*, pages 1797–1807, Brussels, Belgium. Association for Computational Linguistics.

Yurii Nesterov. 1983. A method of solving a convex programming problem with convergence rate $O(1/k^2)$. *Soviet Mathematics Doklady*, 27(2):372–376.

Maria Panagioti, Efharis Panagopoulou, Peter Bower, George Lewith, Evangelos Kontopantelis, Carolyn Chew-Graham, Shoba Dawson, Harm van Marwijk, Keith Geraghty, and Aneez Esmail. 2017. Controlled interventions to reduce burnout in physicians: A systematic review and meta-analysis. *JAMA Internal Medicine*, 177(2):195–205.

Pinal Patel, Disha Davey, Vishal Panchal, and Parth Pathak. 2018a. Annotation of a large clinical entity corpus. In *Proceedings of the 2018 Conference on Empirical Methods in Natural Language Processing*, pages 2033–2042, Brussels, Belgium. Association for Computational Linguistics.

Rikinkumar S Patel, Ramya Bachu, Archana Adikey, Meryem Malik, and Mansi Shah. 2018b. Factors related to physician burnout and its consequences: A review. *Behavioral sciences*, 8(11):98.

Thomas H. Payne, W. David Alonso, J. Andrew Markiel, Kevin Lybarger, and Andrew A. White. 2018. Using voice to create hospital progress notes: Description of a mobile application and supporting system integrated with a commercial electronic health record. *Journal of Biomedical Informatics*, 77:91–96.

Juan C. Quiroz, Liliana Laranjo, Ahmet Baki Kocaballi, Shlomo Berkovsky, Dana Rezazadegan, and Enrico Coiera. 2019. Challenges of developing a digital scribe to reduce clinical documentation burden. *npj Digital Medicine*, 2(1):114.

Abigail See, Peter J. Liu, and Christopher D. Manning. 2017. Get to the point: Summarization with pointer-generator networks. In *Proceedings of the 55th Annual Meeting of the Association for Computational Linguistics (Volume 1: Long Papers)*, pages 1073–1083, Vancouver, Canada. Association for Computational Linguistics.

Sai Prabhakar Pandi Selvaraj and Sandeep Konam. 2020. Medication regimen extraction from clinical conversations. In *International Workshop on Health Intelligence (W3PHIAI 2020)*, New York, USA.

Izhak Shafran, Nan Du, Linh Tran, Amanda Perry, Lauren Keyes, Mark Knichel, Ashley Domin, Lei Huang, Yu hui Chen, Gang Li, Mingqiu Wang, Laurent El Shafey, Hagen Soltau, and Justin Stuart Paul. 2020. The medical scribe: Corpus development and model performance analyses. In *Proceedings of the Twelfth International Conference on Language*

Resources and Evaluation (LREC 2020), Marseille, France. To appear.

Tait D. Shanafelt, Liselotte (Lotte) Dyrbye, Christine Sinsky, Omar Hasan, Daniel Satele, Jeff A. Sloan, and Colin Patrick West. 2016. Relationship between clerical burden and characteristics of the electronic environment with physician burnout and professional satisfaction. *Mayo Clinic Proceedings*, 91(7):836–848.

Peter Shaw, Jakob Uszkoreit, and Ashish Vaswani. 2018. Self-attention with relative position representations. In *Proceedings of the 2018 Conference of the North American Chapter of the Association for Computational Linguistics: Human Language Technologies, Volume 2 (Short Papers)*, pages 464–468, New Orleans, Louisiana. Association for Computational Linguistics.

Christine Sinsky, Lacey Colligan, Ling Li, Mirela Prgomet, Sam Reynolds, Lindsey Goeders, Johanna Westbrook, Michael Tutty, and George Blike. 2016. Allocation of physician time in ambulatory practice: A time and motion study in 4 specialties. *Annals of Internal Medicine*, 165(11):753–760.

Ilya Sutskever, Oriol Vinyals, and Quoc V Le. 2014. Sequence to sequence learning with neural networks. In Z. Ghahramani, M. Welling, C. Cortes, N. D. Lawrence, and K. Q. Weinberger, editors, *Advances in Neural Information Processing Systems 27*, pages 3104–3112. Curran Associates, Inc.

Ashish Vaswani, Noam Shazeer, Niki Parmar, Jakob Uszkoreit, Llion Jones, Aidan N. Gomez, Ł ukasz Kaiser, and Illia Polosukhin. 2017. Attention is all you need. In *Advances in Neural Information Processing Systems 30*, pages 5998–6008. Curran Associates, Inc.

Qiang Wang, Bei Li, Tong Xiao, Jingbo Zhu, Changliang Li, Derek F. Wong, and Lidia S. Chao. 2019. Learning deep Transformer models for machine translation. In *Proceedings of the 57th Annual Meeting of the Association for Computational Linguistics*, pages 1810–1822, Florence, Italy. Association for Computational Linguistics.

Alexi A. Wright and Ingrid T Katz. 2018. Beyond burnout—redesigning care to restore meaning and sanity for physicians. *The New England Journal of Medicine*, 378(4):309–311.

Yuhao Zhang, Daisy Yi Ding, Tianpei Qian, Christopher D. Manning, and Curtis P. Langlotz. 2018. Learning to summarize radiology findings. In *EMNLP 2018 Workshop on Health Text Mining and Information Analysis*.

Towards an ontology-based medication conversational agent for PrEP and PEP

Muhammad (Tuan) Amith, Licong Cui, Kirk Roberts, Cui Tao
School of Biomedical Informatics
The University of Texas Health Science Center at Houston
Houston, Texas

Abstract

HIV (human immunodeficiency virus) can damage a human's immune system and cause Acquired Immunodeficiency Syndrome (AIDS) which could lead to severe outcomes, including death. While HIV infections have decreased over the last decade, there is still a significant population where the infection permeates. PrEP and PEP are two proven preventive measures introduced that involve periodic dosage to stop the onset of HIV infection. However, the adherence rates for this medication is low in part due to the lack of information about the medication. There exist several communication barriers that prevent patient-provider communication from happening. In this work, we present our ontology-based method for automating the communication of this medication that can be deployed for live conversational agents for PrEP and PEP. This method facilitates a model of automated conversation between the machine and user can also answer relevant questions.

1 Introduction

HIV can cause a dangerous infection that can lead to AIDS, a disease that can lead to severe immunological symptoms and eventual death. Common modes of infection include sexual contact, blood transfusion, or the sharing of drug paraphernalia. While the rates have dropped over the last few decades, HIV infection is not uncommon. For example, there is an infection rate of 2 million globally (World Health Organization, 2017) and 39,782 within the United States (Hess et al., 2018). In addition, a segment of the American population with HIV are unaware of the HIV status (Centers for Disease Control and Prevention, 2016), and therefore at risk of spreading the disease to other individuals.

Advances in medication introduced PrEP and PEP. PEP refers to the use of antiretroviral drugs for people who are HIV-negative after a single high-risk exposure to stop HIV infection, while PrEP is a prevention method for people who are HIV-negative and have high risks of HIV infection. Both of these treatments require consistent adherence to the dosage in order to be fully effective, but adherence is an issue for patients subscribed to it. Providers in particular are concerned about the consistent adherence to PrEP (Wood et al., 2018; Blackstock et al., 2017; Clement et al., 2018).

It has been reported that if PrEP adherence is high, rates of HIV infection will be sizeably reduced (Smith et al., 2015). However, adherence to PrEP is no different than other challenges with medications, such as the patient comprehending the administration of the medication and remembering to take it (American Medical Association, 2016). On top of that, the Centers for Disease and Control (CDC) specifically prescribes periodic counseling, and coordinating with patients on a one-on-one basis (Centers for Disease Control and Prevention, 2014). But time burdens and manpower to conduct counseling pose another challenge (Krakower et al., 2014).

In a previous study, Amith et al. (2019a) utilized an ontology-based method to model the dialogue for the counseling of the HPV vaccine. In this study, we tailor the method for PrEP and PEP counseling with the intent that this could be employed in portable tools for drug users to use. A benefit of using an ontology approach, other than exploiting network-based model for dialogue, is the potential to link the ontology to representations of health behavior models (like the transtheoretical model). Systems that leverage health behavior models, according to Kennedy et al. (2012), have demonstrated to be more impactful on affecting health behaviors of users. Also, an ontology that models dialogue can yield standardization and sharing. Amith et al. (2019b) noted from their

Proceedings of the 1st Workshop on NLP for Medical Conversations, pages 31–40
Online, 10, 2020. ©2020 Association for Computational Linguistics
https://doi.org/10.18653/v1/P17

literature review on PubMed that there is limited ontology-centric studies for health-based dialogue management. Amith et al. (2020) simulation studies have also shown evidence that automated counseling, specifically conversational agents for vaccines, could impact the health attitudes and beliefs that can lead to improved uptake with perceived high usability.

Ontologies are artifacts that represent and encode domain knowledge for machines to understand a domain and their physical environment. According to one school of thought, if machines have symbolic understanding of their domain and environment, it could potentially provide near-autonomous operation of tasks. Imbuing software with autonomous task of dialogue interaction requires some measure of intelligence. Intelligent agents are defined as having *reactive, proactive* and *social ability* features (Wooldridge and Jennings, 1995). *Reactive* refers to the software ability for timely response to the environment. *Proactive* refers to the initiative driven aspect of the software to accomplish tasks, and *social ability* involves the software handling external interaction with the environment (virtual or physical). How these qualities manifest vary by the architectural approach (*reactive agents, reasoning agents*, etc.) which is beyond the scope of discussion.

Researchers mention the use of internal data models within the architecture of the agents (Wooldridge, 2009). The models' role in the system is to provide the agent with decision making capabilities to perform autonomously in the environment. This would include 1) representing the domain knowledge for the agent, 2) providing information of the surrounding environment of the agent, and 3) cataloging the previous actions of the agent (e.g., for the agent to learn). According to Hadzic and colleagues, these models could be manifested as a group of ontologies (Hadzic et al., 2009). Furthermore, they state some inherit benefits such as producing shared communication models between agents and systems, information retrieval, organization of the agent's task, and analytical and reasoning of the knowledge (Hadzic et al., 2009).

The ontology-based solution also attempts to solve some of the issues with reasoning agents like the *transduction problem* and the *representation/reasoning problem* (Wooldridge, 2009). The *Transduction problem* is how to translate the world or domain that the agent is embodied in into symbolic representations. The *representation and reasoning problem* pertains to the challenge of manipulating the symbolic representations and applying reasoning for the agent. With ontologies, we can model a domain space or the environment using predicate logic that is syntactically encoded into a machine-readable artifact. Within the context of this work, this method maps utterances of the user and the machine to concepts represented in our ontological model. Also, with the availability of reasoners, like Pellet (Sirin et al., 2007) or HermiT (Glimm et al., 2014; Shearer et al., 2008), we can perform reasoning based on the encoded model to generate inferred dialogue context information.

From a natural language processing (NLP) standpoint, dialogue is essentially a sequence of utterances between multiple agents. Our work utilizes a finite state transition network to model the dialogue (Allen, 1995; Jurafsky and Martin, 2000), and then encodes this sequence model of the utterances within the ontology. We also employ some lightweight NLP methods to help the agent discern participant utterances, alongside with the reasoning capacities of the agent. For the design of the dialogue system, we utilize a deterministic and planned approach to automate the counseling versus a generative approach in order to cover certain main points to communicate to the drug user. This gives us the control needed to ensure the conversational agent delivers the appropriate counseling. The dialogue will center around a closed world domain – specific to only PrEP and PEP, and HIV infection. The following sections will cover the development of the conversational agent and discussion through results of a Trindi Tick assessment for dialogue system evaluation and future steps with our work.

2 Methods

2.1 Ontology Models

We developed a series of ontologies to provide the software agent with interaction abilities – to model patient-level information and the dialogue flow for the agent to coordinate the interaction with the user.

Ontology of PrEP and PEP (OPP) For the PrEP and PEP information source we created the Ontology of PrEP and PEP (OPP), using patient-level sources (brochures and websites). The OPP describes basic dosing, benefit and harms, cost, potential users, and other pertinent information that patients would like to know. This ontology provides a knowledge base for atomic facts for the dialogue flow ontology, PHIDO. This early version of OPP has 152 classes, 57 object properties, 23 data properties, and 10 instance individuals.

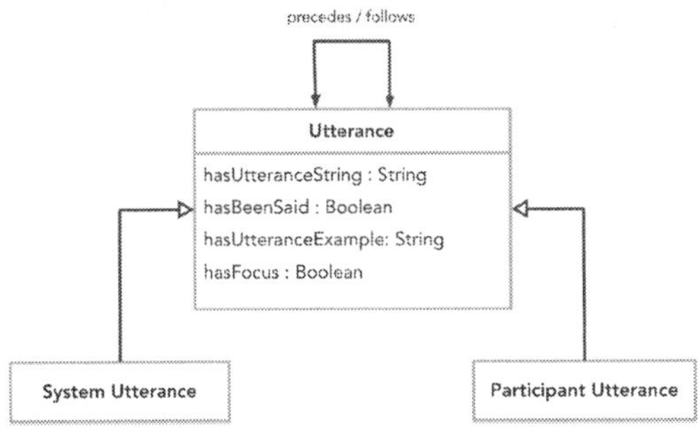

Figure 1: UML diagram of the Utterance class in PHIDO. System and Participant are subclasses of Utterance.

Patient Health Information Ontology Dialogue (PHIDO) The Patient Health Information Ontology Dialogue is an ontology developed in the previous study to model a chain of utterances between the machine (utterances of the system) and the user speaking directly with the machine (utterance of the participant). Figure 1 displays the Utterance class in PHIDO. The parent Utterance class has several data properties that are used to help facilitate the flow the machine's conversation and are linked together using "precedes" or "follows" to indicate precedence of the utterances. PHIDO's TBox level metrics contain 86 classes, 9 object properties, and 5 data properties. Details of the ontology is discussed in the authors' previous study (Amith et al., 2019b).

Essentially, each triple (i.e. predicate) from OPP is utilized by PHIDO to communicate statements about PrEP or PEP (Figure 2). Within PHIDO, an utterance data (instance) is linked to each predicate for the machine to either speak or to help discern utterances spoken by the user.

Figure 3 shows the meta-level description of the dialogue that starts with basic introduction and acclimation of the user with the machine and closing out the counseling. The core goals of the dialogue is to communicate facts (Health Information) and to handle questions at any time for the user (Question Answering). The flow of communication for health information is facilitated by a sub-goal we call Discuss Health Topic (DHT) which is modeled in the PHIDO and allows for population of utterance data that aligns with the concepts in DHT.

2.2 Dialogue System

From a previous study we developed a software engine that uses the aforementioned ontology models (Amith et al., 2019a). The software engine also supports question answering if the ontology model infers a question based on the context of the dialogue flow (e.g., an answer follows a question, a question precedes a point of confusion in the conversation, etc.).

On Figure 4, we demonstrate the execution of the engine using the PHIDO to model the communication of patient-level information to the user. Here the system evokes each patient-level health fact (Health Information) about PrEP, sourced from the OPP. Any question recognized by the engine will attempt to answer the question using an ontology-based question answering (QA) sub-system of the engine. Once the utterance of the user is identified as a question, the type of question, the nouns, and verb phrases are analyzed against the OPP predicates. After selection, ranking, and filtering, the answer is evoked by the system. Details about the implementation is described in (Amith et al., 2019a), and Figure 5 shows an example demonstration with a question and a response from the QA system.

2.2.1 NLP-based Slave Functions

Aside from the finite-state diagram approach for the engine, we also employed some NLP methods as slave functions for the engine to operate - discerning the type of participant utterance for the main dialogue system and comparing the question data with ontology triples for the question-answering subsystem.

Discerning Participant Utterances When capturing input from the user, the engine will need to distinguish the type of user utterance to direct the flow of the dialogue. Each utterance concept was annotated to a set of examples that were representative of its type. For example, the Utterance class of Question has string examples of "could you tell me", "how many", "list", and the Disconfirmation class has examples such as "negative", "never", "none", etc. Using the inputted utterance of the user, the text is compared to the string examples using Monge-Elkan (Monge et al., 1996;

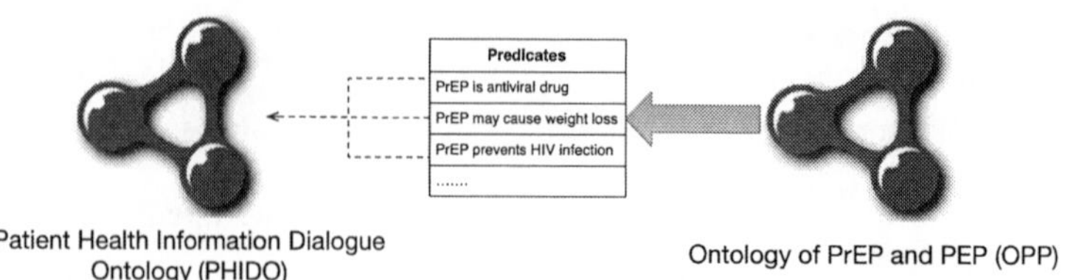

Figure 2: Triples (Predicates) are extracted from the Ontology of PrEP and PEP and used by PHIDO.

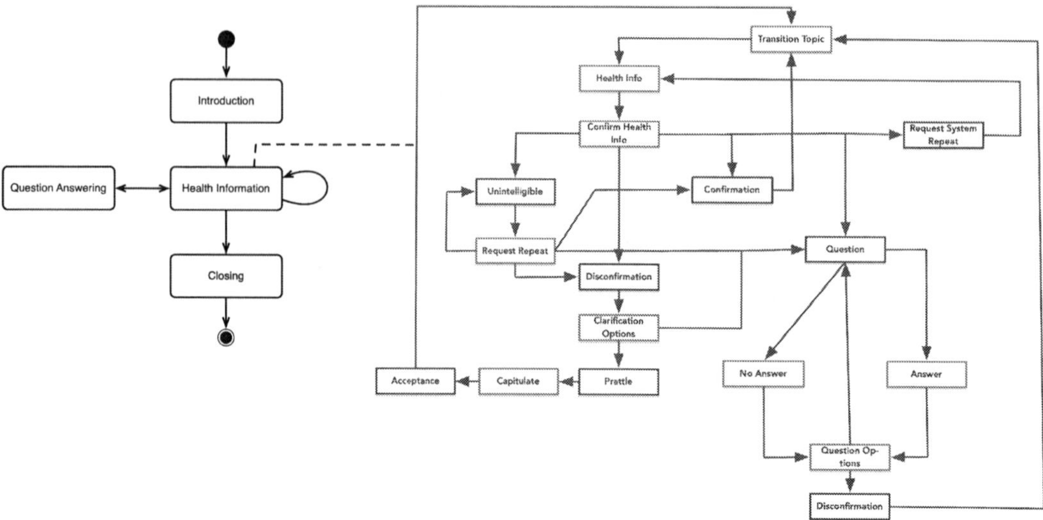

Figure 3: Dialogue flow plan for basic counseling for PrEP and PEP information.

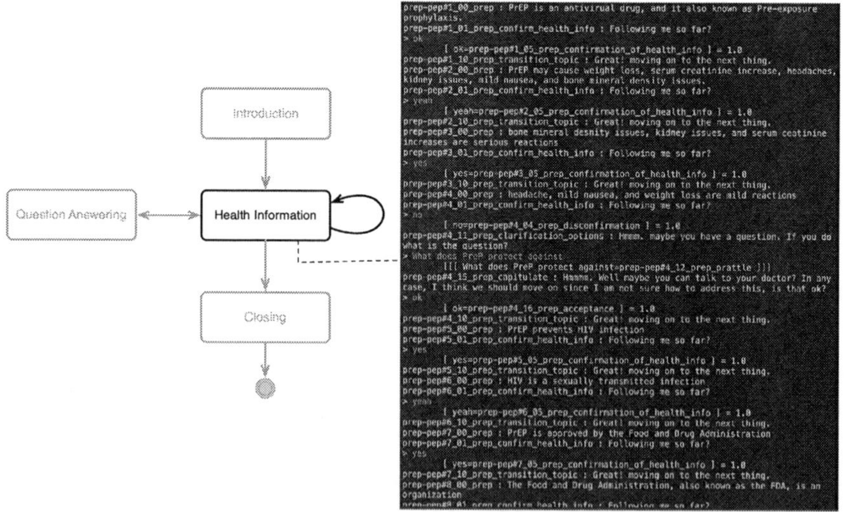

Figure 4: Demonstration of the dialogue engine communicating PrEP information to the user.

Monge and Elkan, 1997) (default implementation from Korstanje (2019)), and a default threshold of 0.85. Matches that do not meet the threshold will fallback to exact string matches based on the beginning of the string.

Definition 2.1 (Participant utterance & examples).

Every participant utterance PU expected by the system contains example data EU_n. EU_n has a number of string text TT_n that are a set of tokens

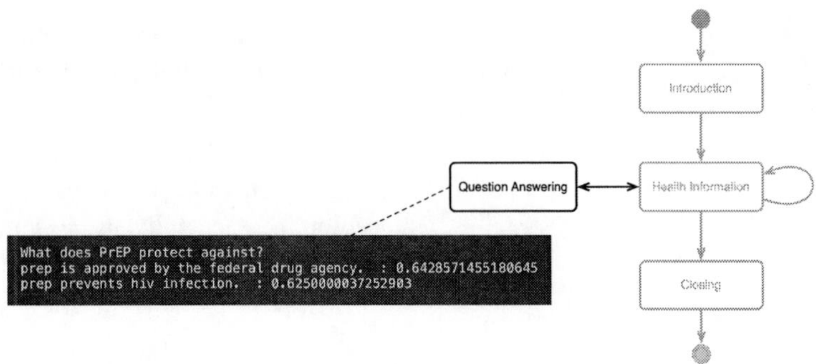

Figure 5: Demonstration of the a PrEP-related question asked with corresponding answers provided. The ranking scores are provided for supplement.

t_n.

$$\forall\, PU^n \ni EU_n^n$$

$$EU_n^n \in \begin{cases} TT_n, where \\ TT = \{t_1 t_2 \cdots t_n\} \\ t \ is \ a \ string \ token \end{cases}$$

Definition 2.2 (User utterance). *User utterance UU for the dialogue system is a set of string tokens t_n*

$$\forall\, UU = \{t_1 t_2 \cdots t_n\}$$

Definition 2.3 (Identifying the participant utterance). *To find the exactly identified PU^n the dialogue system attains a comparison $s(x)$ that is the maximum and the greater than a defined threshold TH among all of the example utterances EU^n within each expected utterance PU^n.*

$$PU^n \Rightarrow T_{score} = max(s(x)) > TH$$

$$s(x) = \begin{cases} EU_n^n \cap UU, \\ where \\ EU_n^n = t_1^e t_2^e \cdots t_n^e \end{cases}$$

Defintion 2.3.1 (Identifying the participant utterance). *Assuming that Definition 2.3 fails to find the expected participant utterance PU^n, the dialogue system resorts to finding exact match of the beginning string tokens EU' and UU' from example utterances EU_n^n and the user utterance UU.*

$$PU^n \Rightarrow T_{score} = b(x)$$

$$b(x) = \{EU_n^n \approx UU \Rightarrow EU_n' = UU'\},$$

$$where$$

$$EU_n' \in EU_n^n = \begin{cases} EU_n' = \{t_1^e t_2^e \cdots t_{n-m}^e\} \\ EU_n^n = \{t_1^e t_2^e \cdots t_n^e\} \end{cases}$$

$$UU' \in UU = \begin{cases} UU' = \{t_1 t_2 \cdots t_{n-m}\} \\ UU = \{t_1 t_2 \cdots t_n\} \end{cases}$$

Comparing Question Data and Ontology Triples For the question answering subsystem, the system utilized off the shelf NLP tools like Stanford Core (Manning et al., 2014) to extract data from the question. To preform the matching described in (Amith et al., 2019a), we also utilized a combination of either word embedding using Numberbatch vector model (Speer and Lowry-Duda, 2017) (with Semantic Vectors (Widdows and Cohen, 2010) as the interface layer) or the string similarity methods discussed earlier, and extJWNL (Autayeu, 2016) where we assign a score to each triple from the knowledge base (OPP). For brevity, we applied various rules and thresholds to select and filter triples to present an answer.

Definition 2.4 (Primary Question Data). *Given a question Q, there are a subset of elements NP_n and VP_n (noun phrases and verb phrases) that are essential data Q' for the subsystem.*

$$Q' \in \{NP_n, VP_n\}$$

Definition 2.5 (Triple Assertion). *Within an ontology O, there are assertion triples (ABox) that are composed of elements of subject s_n, predicate p_n, and object o_n to form an assertion triple spo_n.*

$$spo = \{s, p, o\}$$

Definition 2.6 (Essential Ontology Triples). *Given an target ontology O, there are a subset of triple assertion types spo^o, spo^d, spo^c (object property assertions, data property assertions, and class assertions) that are needed O' for the subsystem.*

$$O' \in \{spo_n^o, spo_n^d, spo_n^c\}$$

Definition 2.7 (Assign Score From Comparison). *A similarity score TS is assigned from comparing* *the similarity of question data D' with a triple assertion $spo_n^{\{o,d,c\}}$ from O'. TS is derived from the mean of computing similarity between NP_n with $s_n^{\{o,d,c\}}$ and $o_n^{\{o,d,c\}}$, and from VP_n with $p_n^{\{o,d,c\}}$.*

$$Q' \approx spo_n^{\{o,d,c\}} := TS$$

$$TS \begin{cases} sim(s_n^{\{o,d,c\}}, NP_n) = S_n^{sn} \\ sim(o_n^{\{o,d,c\}}, NP_n) = S_n^{on} \\ sim(p_n^{\{o,d,c\}}, VP_n) = S_n^{pv} \\ TS = mean(S_n^{sn}, S_n^{on}, S_n^{pv}) \end{cases}$$

3   Results

We utilized the Trindi Tick list to ascertain the comprehensive functionality of the dialogue engine's ability (Bickmore et al., 2018; Bos et al., 1999). With the exception for the classic Trindi Tick list, according to Bickmore et al. (2018), there is limited formalized standard evaluations for dialogue systems. The Trindi Tick list covers 12 criteria pertaining to handling quality and quantity of user responses, sub-dialogue flows, intelligible responses, etc.

3.1   Is utterance interpretation sensitive to context

No, specifically pertaining to personal contextual information of the user. For example, personal contextual information that is tailored to specific demographic based on behavior or demographic. Another case is a repeat of information that has already been resolved. For example, a user asked a question about side effects of PrEP and then later on the system communicated side effects of PrEP. Ideally it would preferable to skip over the side effect of PrEP if the user had asked the question already.

3.2   Can the system deal with answers to questions that give more information than was requested

Partially yes. This criterion is based on the coverage of the ontology, as the system will attempt to match every predicate that is associated with entities and relationship from a user question.

3.3   Can the system deal with answers to questions that give different information than was actually requested

and

3.4   Can the system deal with answers to questions that give less information than was requested?

Yes, for example, a user can ask a question at any point which will activate the question-answering subroutine. Generally, the system overall is indiscriminate about the amount of information said by the user and will react.

3.5   Can the system deal with ambiguous designators

Partially yes. Similar to criterion 3.2, this is based on the coverage of the ontology and the similarity measurements.

3.6   Can the system deal with negatively specified information

No, not intentionally. We presume that this criterion refers to non-explicit negative responses that compliment the designators, other than stating no or similar negative utterances. With the former, the negated tokens are ignored when the system analyzes the utterance. With the latter, if expected, the system can deal with those responses (e.g., "nah", "negatory", etc.).

> **System:***PrEP may cause weight loss, serum creatinine increase, headaches, kidney issues, mild nausea, and bone mineral density issues. Following me so far?*
> **Participant**: *nope*
> **System**: *Hmmm. maybe you have a question. If you do what is the question?*

3.7   Can the system deal with no answer to a question at all

No. The system will wait for a response by the user.

3.8   Can the system deal with noisy input

Yes, by "noisy" we refer to utterances that might be indistinguishable if mis-transcribed or garbled text input.

> **System:** *Great! moving on to the next thing. PrEP may cause weight loss, serum creatinine increase, headaches, kidney issues, mild nausea, and bone mineral density issues. Following me so far?*
> **Participant:** *super calif ragilistic expialido cious*

System: *I am sorry. would you mind repeating what you said?*

Participant: *Yes, I follow.*

System: *Great! moving on to the next thing...*

3.9 Can the system deal with "help" sub-dialogs initiated by the user

Yes. A relevant "help" example is when the system facilitates any repeat of information.

System: *headache, mild nausea, and weight loss are mild reactions. Following me so far?*

Participant: *say that again*

System: *headache, mild nausea, and weight loss are mild reactions. Following me so far?*

3.10 Can the system deal with "non-help" sub-dialogs initiated by the user

Yes, a "non-help" sub-dialogue example in our system is whenever a user has additional question that is related to the topic discussed.

System: *PrEP is an antivirual drug, and it also known as Pre-exposure prophylaxis. Following me so far?*

User: *What are some payment options to cover PrEP?*

System: *That is a great question. I would recommend discussing it with your doctor for confirmation. If you have another question, I can help you further. Ask away.* (System did not have an answer and provides one of the default responses.)

User: *no further questions*

System: *Great! moving on to the next thing. PrEP may cause weight loss, serum creatinine increase, headaches, kidney issues, mild nausea, and bone mineral density issues. Following me?*

3.11 Does the system only ask appropriate follow-up questions

Yes. After each segment, the system inquires to the user if they have any follow-up responses to what they heard about PrEP. Example in 3.10 demonstrates this.

3.12 Can the system deal with inconsistent information

Partially yes. The system relies on example of expected utterances to identify the type of utterance using string metric similarity. This may result in misidentifying the utterance and directing the dialogue flow in unintended direction.

4 Discussion

The apparent limitations of the system is highlighted by criterion 3.1, 3.6, and 3.7. The limitation with respect to context is primarily due to lack of a mechanism to handle personalized information. One of the benefits of using ontologies demonstrated by health researchers was the potential to tailor information if we were to capture user information (Bickmore et al., 2011). Previous studies have demonstrated the use of user context ontologies to reason with user data. We assume that this component could be integrated to support personalized contextual information based on group identification or past previous behavior of the user.

Another limitation is the negatively specified information, where if a user were to ask "What if I do not have insurance to pay for PrEP?" Technically the system would not analyze the negative token "not" and focus on the more salient entities of the response. However, a response can be generated by the system, but whether it would accurately respond to the question is unknown, and is determined by the scope of the ontology.

In regards to dealing with no answer, the system awaits for the response of the participant. The reasonable solution is to implement a software code subroutine either on the dialogue system level or on the interface level that times out whenever the user does not provide a timely response. Nonetheless, exploring how this can be done on the ontology-level would need to be investigated and engineered into the ontology.

Other aspects highlighted by our preliminary Trindi Tick assessment underline adherence to criteria regarding handling indistinguishable responses, sub-dialogue branches, and the quality and quantity of information. Also, the quality of the system responsiveness, we theorize, would be dependent on the scope of the knowledge encoded in the ontology.

5 Conclusion

In this paper we present our ontology-based system for handling dialogue for PrEP and PEP counseling. This system also handles questions that are queried from a knowledge base, called the Ontology of PrEP and PEP (OPP). Overall the objective of this work is to demonstrate the feasibility of using an ontology-driven approach to manage automated counseling for PrEP and PEP through a computer-based agent.

Figure 6 shows overall deployment on how the engine will interface with external natural language clients whether they are mobile or terminals (desktops or kiosks). Our eventual goal is to develop a deterministic, planned-based approach within the domain of PrEP and PEP medication adherence (closed domain) and test our approach with live participants.

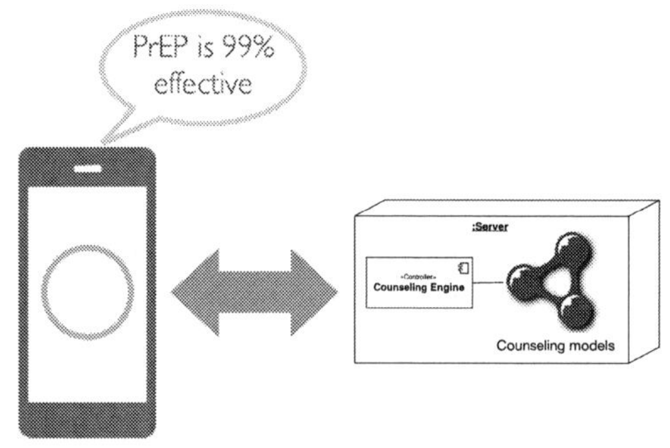

Figure 6: Deployment of a conversational agent client as a mobile application.

Previous research (Amith et al., 2019b) has found limited use of ontologies for medical-based dialogue agents. Solutions addressing PrEP or PEP adherence that researchers have examined include social networks (Kuhns et al., 2017; Garcia et al., 2016) and telehealth solutions (Klausner and CFAR Development Core, 2018; Stekler et al., 2018; Youth Tech Health, 2018). With the former, there has not been any evidence that shows that social networks can address adherence or awareness (Ezennia et al., 2019), and with the latter, telehealth solutions are limited to the availability of a professional and may not be cost effective (Touger and Wood, 2019). Having an automated agent that can provide real-time and high availability to counsel and inform patients may offer an alternative, but further research is needed to foresee this possibility.

Limitations and Future Direction The ontologies that drive the system are currently in draft format and additional work is needed to expand them to include more personalized content, such as where PrEP and PEP can be obtained and information for nonprofit organizations that can provide support, etc. Researchers have conducted simulations to fine tune a formal plan to counsel individuals on the HPV vaccine. Our future work would need to model standard practices for medication counseling adherence that typically happen between patients and providers. This would include conducting simulation studies and working with providers to develop, and then model the counseling flow using the PHIDO framework. Also, from the sample dialogue of the simulation, we can parse out potential questions that can be used to test the question answering component. Lastly, the demonstration of our work is based on text-based modality, and we are working towards interfacing the system to a voice interface to capture the user utterances and evoke the utterance of the machine.

Acknowledgments

Research was supported by the National Library of Medicine of the National Institutes of Health under Award Numbers R01LM011829 and R00LM012104, and the National Institute of Allergy and Infectious Diseases of the National Institutes of Health under Award Number R01AI130460.

References

James Allen. 1995. *Natural language understanding.* Pearson.

American Medical Association. 2016. 8 reasons patients don't take their medications.

Muhammad Amith, Rebecca Lin, Licong Cui, Dennis Wang, Anna Zhu, Grace Xiong, Hua Xu, Kirk Roberts, and Cui Tao. 2019a. An ontology-powered dialogue engine for patient communication of vaccines. In *4th International Workshop on Semantics-Powered Data Mining and Analytics (SEPDA 2019) in Conjunction with the 19th International Semantic Web Conference (ISWC 2019).*

Muhammad Amith, Rebecca Lin, Rachel Cunningham, Qiwei Luna Wu, Lara S Savas, Yang Gong, Julie A Boom, Lu Tang, and Cui Tao. 2020. Examining potential usability and health beliefs among young adults using a conversational agent for hpv vaccine counseling. In *2020 American Medical Informatics Association Informatics Summit.*

Muhammad Amith, Kirk Roberts, and Cui Tao. 2019b. Conceiving an application ontology to model patient human papillomavirus vaccine counseling for dialogue management. *BMC bioinformatics*, 20(21):1–16.

Aliaksandr Autayeu. 2016. extJWNL.

Timothy Bickmore, Ha Trinh, Reza Asadi, and Stefan Olafsson. 2018. Safety first: Conversational agents for health care. In *Studies in Conversational UX Design*, pages 33–57. Springer.

Timothy W Bickmore, Daniel Schulman, and Candace L Sidner. 2011. A reusable framework for health counseling dialogue systems based on a behavioral medicine ontology. *Journal of biomedical informatics*, 44(2):183–197.

Oni J Blackstock, Brent A Moore, Gail V Berkenblit, Sarah K Calabrese, Chinazo O Cunningham, David A Fiellin, Viraj V Patel, Karran A Phillips, Jeanette M Tetrault, and Minesh Shah. 2017. A cross-sectional online survey of hiv pre-exposure prophylaxis adoption among primary care physicians. *Journal of general internal medicine*, 32(1):62–70.

Johan Bos, Staffan Larsson, I Lewin, C Matheson, and D Milward. 1999. Survey of existing interactive systems. *Trindi (Task Oriented Instructional Dialogue) report*, (D1):3.

Centers for Disease Control and Prevention. 2014. Pre-exposure prophylaxis for the prevention of HIV infection in the United States–2014: A clinical practice guideline.

Centers for Disease Control and Prevention. 2016. Monitoring selected national hiv prevention and care objectives by using hiv surveillance data—united states and 6 dependent areas, 2014. Report, Centers for Disease Control and Prevention.

Meredith E Clement, Jessica Seidelman, Jiewei Wu, Kareem Alexis, Kara McGee, N Lance Okeke, Gregory Samsa, and Mehri McKellar. 2018. An educational initiative in response to identified prep prescribing needs among pcps in the southern us. *AIDS care*, 30(5):650–655.

Ogochukwu Ezennia, Angelica Geter, and Dawn K Smith. 2019. The prep care continuum and black men who have sex with men: A scoping review of published data on awareness, uptake, adherence, and retention in prep care. *AIDS and Behavior*, 23(10):2654–2673.

Jonathan Garcia, Caroline Parker, Richard G Parker, Patrick A Wilson, Morgan Philbin, and Jennifer S Hirsch. 2016. Psychosocial implications of homophobia and hiv stigma in social support networks: insights for high-impact hiv prevention among black men who have sex with men. *Health Education and Behavior*, 43(2):217–225.

Birte Glimm, Ian Horrocks, Boris Motik, Giorgos Stoilos, and Zhe Wang. 2014. Hermit: an owl 2 reasoner. *Journal of Automated Reasoning*, 53(3):245–269.

Maja Hadzic, Pornpit Wongthongtham, Tharam Dillon, and Elizabeth Chang. 2009. *Ontology-based multi-agent systems*. Springer.

Kristen L Hess, Shacara D Johnson, Xiaohong Hu, Jianmin Li, Baohua Wu, Chenchen Yu, Hong Zhu, Chang Jin, Mi Chen, and John Gerstle. 2018. Diagnoses of hiv infection in the united states and dependent areas, 2017. *HIV Surveillance Report*.

Daniel Jurafsky and James Martin. 2000. *Speech and Language Processing: An Introduction to Natural Language Processing, Computational Linguistics, and Speech Recognition*, second edition edition. Pearson Prentice Hall, Upper Saddle River, NJ.

Catriona M Kennedy, John Powell, Thomas H Payne, John Ainsworth, Alan Boyd, and Iain Buchan. 2012. Active assistance technology for health-related behavior change: an interdisciplinary review. *Journal of medical Internet research*, 14(3):e80.

J Klausner and CFAR Development Core. 2018. [link].

M.P. Korstanje. 2019. SimMetrics.

Douglas Krakower, Norma Ware, Jennifer A Mitty, Kevin Maloney, and Kenneth H Mayer. 2014. Hiv providers' perceived barriers and facilitators to implementing pre-exposure prophylaxis in care settings: a qualitative study. *AIDS and Behavior*, 18(9):1712–1721.

Lisa M Kuhns, Anna L Hotton, John Schneider, Robert Garofalo, and Kayo Fujimoto. 2017. Use of pre-exposure prophylaxis (prep) in young men who have sex with men is associated with race, sexual risk behavior and peer network size. *AIDS and Behavior*, 21(5):1376–1382.

Christopher D Manning, Mihai Surdeanu, John Bauer, Jenny Rose Finkel, Steven Bethard, and David McClosky. 2014. The stanford corenlp natural language processing toolkit. In *Proceedings of 52nd annual meeting of the association for computational linguistics: system demonstrations*, pages 55–60.

Alvaro Monge and Charles Elkan. 1997. An efficient domain-independent algorithm for detecting approximately duplicate database records. *Proc. of the ACM-SIGMOD Workshop on Research Issues on Knowledge Discovery and Data Mining*.

Alvaro E Monge, Charles Elkan, et al. 1996. The field matching problem: Algorithms and applications. In *Kdd*, volume 2, pages 267–270.

Rob Shearer, Boris Motik, and Ian Horrocks. 2008. Hermit: A highly-efficient owl reasoner. In *Owled*, volume 432, page 91.

Evren Sirin, Bijan Parsia, Bernardo Cuenca Grau, Aditya Kalyanpur, and Yarden Katz. 2007. Pellet: A practical owl-dl reasoner. *Journal of Web Semantics*, 5(2):51–53.

Dawn K Smith, Jeffrey H Herbst, and Charles E Rose. 2015. Estimating hiv protective effects of method adherence with combinations of preexposure prophylaxis and condom use among african american men who have sex with men. *Sexually transmitted diseases*, 42(2):88–92.

Robert Speer and Joanna Lowry-Duda. 2017. Conceptnet at semeval-2017 task 2: Extending word embeddings with multilingual relational knowledge. *arXiv preprint arXiv:1704.03560*.

Joanne D Stekler, Vanessa McMahan, Lark Ballinger, Luis Viquez, Fred Swanson, Jon Stockton, Beth Crutsinger-Perry, David Kern, and John D Scott. 2018. Hiv pre-exposure prophylaxis prescribing through telehealth. *JAIDS Journal of Acquired Immune Deficiency Syndromes*, 77(5):e40–e42.

Rebecca Touger and Brian R Wood. 2019. A review of telehealth innovations for hiv pre-exposure prophylaxis (prep). *Current HIV/AIDS Reports*, 16(1):113–119.

Dominic Widdows and Trevor Cohen. 2010. The semantic vectors package: New algorithms and public tools for distributional semantics. In *2010 IEEE Fourth International Conference on Semantic Computing*, pages 9–15. IEEE.

Brian R Wood, Vanessa M McMahan, Kelly Naismith, Jonathan B Stockton, Lori A Delaney, and Joanne D Stekler. 2018. Knowledge, practices, and barriers to hiv preexposure prophylaxis prescribing among washington state medical providers. *Sexually transmitted diseases*, 45(7):452–458.

Michael Wooldridge. 2009. *An introduction to multiagent systems*. John Wiley and Sons.

Michael Wooldridge and NR Jennings. 1995. Intelligent agents: Theory and practice the knowledge engineering review. *The knowledge engineering review*, 10:115–152.

World Health Organization. 2017. World health statistics 2017: Monitoring health for the sdgs.

Youth Tech Health. 2018. [link].

Heart Failure Education of African American and Hispanic/Latino Patients: Data Collection and Analysis

Itika Gupta,[1] Barbara Di Eugenio,[1] Devika Salunke,[2] Andrew D. Boyd,[2]
Paula Allen-Meares,[3] Carolyn A. Dickens,[3] Olga Garcia-Bedoya[3]
[1]Department of Computer Science
[2]Department of Biomedical and Health Information Sciences
[3]Department of Medicine
University of Illinois at Chicago, Chicago, Illinois
{igupta5,bdieugen,dsalun2,boyda}@uic.edu
{pameares,cdickens,ogarciab}@uic.edu

Abstract

Heart failure is a global epidemic with debilitating effects. People with heart failure need to actively participate in home self-care regimens to maintain good health. However, these regimens are not as effective as they could be and are influenced by a variety of factors. Patients from minority communities like African American (AA) and Hispanic/Latino (H/L), often have poor outcomes compared to the average Caucasian population. In this paper, we lay the groundwork to develop an interactive dialogue agent that can assist AA and H/L patients in a culturally sensitive and linguistically accurate manner with their heart health care needs. This will be achieved by extracting relevant educational concepts from the interactions between health educators and patients. Thus far we have recorded and transcribed 20 such interactions. In this paper, we describe our data collection process, thematic and initiative analysis of the interactions, and outline our future steps.

1 Introduction

Heart failure (HF) is defined as "a complex clinical syndrome that can result from any structural or functional cardiac disorder that impairs the ability of the ventricle to fill or eject blood" (Hunt et al., 2009). Approximately 5 million Americans currently live with this condition. In the United States, minority communities have a higher mortality rate than Caucasians (Roger, 2013; Toukhsati et al., 2019). This has been attributed to multiple factors like genetic variations, access to quality healthcare, socioeconomic conditions, health behavior, lower health literacy among others. However, some of these risk factors can be mitigated (Der Ananian et al., 2018; Tucker et al., 2011). For example, a patient with access to personalized educational material is better equipped to identify and address his self-care needs resulting in increased compliance

and better health outcomes (Alberti and Nannini, 2013).

Self-care is "a naturalistic decision-making process by which individuals make choices about behaviors that maintain physiologic stability and the response to symptoms when they occur." (Riegel et al., 2004) However, this process can be rendered ineffective when the patient has a limited understanding of the disease. Furthermore, most self-care materials available outside the hospital are catered towards the White Caucasian educated population, and thus lack cultural nuances to assist patients from minority communities (Barrett et al., 2019; Hughes and Granger, 2014; Lee et al., 2011). This has resulted in poor heart self-care regimen in minority communities (Howie-Esquivel, 2014).

Therefore, we intend to develop a dialogue agent that can provide medically reliable and culturally sensitive self-care information to discharged African American and Hispanic/Latino HF patients, and help mitigate the health disparities observed among them. In this paper, we talk about our first step towards building the agent i.e. collecting the data (since there is no publicly available dataset) and analyzing it. We used topic modeling to identify core educational concepts and analyzed the data for initiative, i.e., who takes the conversational lead. Not surprisingly, educators take more initiative, however the portions in which the patient has control are more important to uncover what patients may ask of a dialogue agent.

We also tried to evaluate the interactions for cultural competency. However, existing tools such as Cross-cultural counseling inventory (LaFromboise et al., 1991) and the Multicultural counseling inventory (Sodowsky et al., 1994) focus on provider's knowledge and do not evaluate patient educational materials. Therefore, with the help of content experts in our team, fundamental concepts of cross-cultural care (empathy, respect, and curiosity), and

41

Proceedings of the 1st Workshop on NLP for Medical Conversations, pages 41–46
Online, 10, 2020. ©2020 Association for Computational Linguistics
https://doi.org/10.18653/v1/P17

focus groups (Bobo et al., 1991; González-Lee and Simon, 1987), we will manually identify culturally relevant topics and model the dialogue agent accordingly as part of our future work.

2 Related Work

In the 1960s, ELIZA was the first Natural Language Processing (NLP) based chatbot which facilitated a dialogue between humans and machines. Since then, multiple advances have been made in artificial intelligence and NLP resulting in the evolution of dialogue agents. They have transitioned from accepting very restricted user input (answers to multiple-choice questions only) to processing full sentences and providing medically reliable information (Laranjo et al., 2018).

Multiple randomized control trials have established the efficacy of dialogue agents in healthcare settings as well (Bickmore et al., 2013a,b; Lovell et al., 2017). They have been successfully used to promote a healthy lifestyle, increase adherence, or provide adjunct psychotherapy among other uses (Laranjo et al., 2018). Technology-based interventions have been used to assist HF failure patients for quite some time. Most of these interventions are catered towards remote monitoring and medication management (Hughes and Granger, 2014). CARDIAC (Computer Assistant for Robust Dialogue Interaction and Care), a conversation assistant for chronic HF patients was designed to collect both objective and subjective information from the patients (Ferguson et al., 2009). Similarly, DIL, another conversation agent was designed to help HF patients to transition from hospital to their homes by motivating them to follow a healthy lifestyle and maintaining medication adherence (Moulik, 2019). To our knowledge, there is no existing culturally sensitive dialogue agent designed to assist minority communities with their heart failure self-care needs.

3 Data Collection

We recruited three health educators to conduct HF education of AA and H/L patients in both the inpatient and outpatient clinics of our university. We plan to collect 40 HF education sessions, half with AA patients and half with H/L patients. We have recorded 20 interactions so far, 18 with AA patients and two with H/L patients; of these 20 patients, 8 are males and 12 females. One of the barriers to recruiting H/L patients is how our hospital iden-

---

> **Patient**: Yeah, I don't, I don't do the frozen meal.
> **Educator**: Okay.
> **Patient**: I was basically doing the uh, vegetables.
> **Educator**: Okay.
> **Patient**: Frozen vegetables,
> **Educator**: They should be fine.
> **Patient**: Yeah.
> **Educator**: But... but, but I do want you to start looking at those nutrition labels.
> **Patient**: Okay.
> **Educator**: And look for something that says less than 5%.
> **Patient**: Okay.
> **Educator**: So, the other thing we always want you to do is, um, of course take all your medicines like you're supposed to.
> **Patient**: Which I didn't do last night.
> **Educator**: Okay.

Figure 1: Excerpt from a conversation

20 transcripts			
	turns	**sentences**	**words**
Educator	116.90	205.45	2281.10
Patient	108.40	131.20	849.50
Total	225.30	336.65	3130.60

Table 1: Distributional analysis of the conversations

tifies them; additionally, since at the moment we focus on English as the language of interaction, we exclude H/L patients if the interaction is conducted in Spanish. Lastly, H/L patients comprise only 20% of our hospital population; this is less than half of the AA patients (45%). The remaining 35% comprises 10% Asian American, and 25% Caucasian and others.

All the 20 recordings were transcribed by a professional transcription service. An excerpt is shown in Figure 1. We should note that in some cases, a third person (a caregiver, like a spouse) is present, and the conversation may involve both patient and caregiver, or be mostly between the educator and the caregiver. While transcribers did a good job, they failed to capture linguistic practices and choices of patients (vernacular speech) and converted it to standard English: for example, 'gonna' was transcribed as 'going to'. Given our focus is on cultural sensitivity, such linguistic practices are of great importance to us, and therefore, the transcripts were revised again to make sure that exactly what was said is recorded.

The average length of an interaction is about 15 minutes. Table 1 presents the average number of turns, sentences, and words per conversation over these 20 HF education sessions. A turn refers to a complete unit of speech and can consist of multiple sentences. Therefore, it makes sense that

Figure 2: Word cloud of common nouns: both educators and patients, the patients, and the educators (left to right)

	Interpretation	Example words from top 10
1	low salt diet	food, sodium, label, meat
2	symptoms	pill, fluid, weight, swell
3	explaining heart failure	pump, failure, blood, body
4	medication and follow-up appointment	appointment, medication, hospital, doctor

Table 2: Interpretation of LDA topics

patients	educators
eating, cooking, restaurant, health, liquid, ocean, sailing, water, **children**, speaking, **negative_emotion**, medical_emergency, **shopping**, **party**, communication (same count as healing)	eating, health, cooking, liquid, sailing, ocean, water, speaking, restaurant, communication, healing, **giving**, medical_emergency, **business**, **cleaning**

Table 3: Most common categories in order of frequency

both the educator and patient had a similar number of turns except the cases where the caregiver did some of the talking. However, there is a noticeable difference in the number of sentences and words between the two. The educator spoke 2.5 times more words than the patient. This indicates that the educator did most of the talking during the interviews. Figure 2 shows the most common nouns in the conversations as word clouds. The frequency of words provides some indication of the topics of conversation, but we now turn to a deeper analysis of these interactions.

4 Data Analysis

In this section, we will look at two types of analysis: thematic analysis and initiative analysis.

Thematic analysis looks at the most common topics of discussion related to HF education. The educators aimed to provide comprehensive information about HF to the patients so that they can take care of themselves upon discharge. The educators covered topics such as what HF is, the role of medications such as the "water pill" (a diuretic), the importance of a low sodium diet, the benefits of physical activity, daily checks to recognize symptoms,

and the value of follow-up appointments. The topics were motivated by the standard HF education information that should be given to a HF patient. However, it is the discussion that arises from these topics that varies based on culture and even person to person. Therefore, we used existing topic modeling tools to further analyze the transcripts.

We first performed topic modeling using Latent Dirichlet Allocation (LDA) (Blei et al., 2003). We used the Gensim library consisting of python wrapper for LDA from MALLET, the java topic modeling toolkit. We tried topic modeling for both the educator and the patient turns together and separately. Since the educators did most of the talking, and almost 50% of patient turns involved filler sentences such as 'okay', 'umm', the patient turns in themselves weren't sufficient to learn separate topics. The remaining 50% of patient turns focused mostly on salt/food which was one of the recurring topic words during the experiments. Therefore, we decided to model the educator turns exclusively. After experimenting with a different number of topics and alpha parameters (Dirichlet prior on the per-document topic distributions), we found the coherence score was the highest for topics counts = 4 and alpha = 0.01. We show the output in Table 2.

We then used Empath to identify the difference in topics between the patients and the educators (Fast et al., 2016). Empath is a text analysis tool that can help identify various topical and emotional categories present in a text. It consists of 200 built-in categories and allows to create more categories on demand. We analyzed both educator and patient turns separately using the built-in categories and show the topmost common 15 categories in Table 3. We used the raw counts for categories in empath. The categories in bold indicate categories different between the educators and the patients. One can notice that 11 of the categories are the same and the most common is *eating*. This is consistent with the word clouds where salt and food were two of the most common nouns.

However, it is interesting to notice that patients

```
_._.._._.._ Control: Patient _._.._._.._
Patient: Yeah, I don't, I don't do the frozen meal. (asser-
tion/command)
Educator: Okay. (prompt)
Patient: I was basically doing the uh, vegetables. (asser-
tion/command)
Educator: Okay. (prompt)
Patient: Frozen vegetables, (assertion/command)
_._.._._.._ Control: Educator _._.._._.._
Educator: They should be fine. (assertion/command)
Patient: Yeah. (prompt)
Educator: But... but, but I do want you to start looking at
those nutrition labels. (assertion/command)
Patient: Okay. (prompt)
Educator: And look for something that says less than 5%.
(assertion/command)
Patient: Okay. (prompt)
Educator: So, the other thing we always want you to do is,
um, of course take all your medicines like you're supposed
to. (assertion/command)
_._.._._.._ Control: Patient _._.._._.._
Patient: Which I didn't do last night. (assertion/command)
Educator: Okay. (prompt)
```

Figure 3: Example conversation showing utterance type and control transfer.

prioritized *children* in their discussion (*family* was also in the top 20 categories), whereas for educators neither of them were even in the top 25 categories. *negative_emotion*, *shopping*, and *party* were in the top 20 of educator categories, therefore can be considered similar to patient categories where they are in top 15. Lastly, *giving*, *business*, and *cleaning* categories were more common in educators. This is because *giving* relates to the term 'give' which educators used frequently to provide information such as 'give you a followup appointment', 'give you a phone number to call', 'give you medicine'. *Business* relates to terms such as 'need', 'work', and 'company', which, similarly to *giving*, was used to inform patients about different companies offering low sodium salt, what they need to do upon discharge, and to educate them about how medications work. *Cleaning* is in the top categories because it relates to the term 'water' which can be considered a partial duplicate of category *water*.

Initiative analysis focuses on the distribution of turns based on the person taking the lead in the conversation. A person takes the lead/initiative when he/she contributes to the conversation (e.g., by asking a question) instead of only answering the questions or responding with fillers (such as 'okay', 'umm'). In turn, when a speaker takes initiative, the control of the conversation transfers to that speaker and remains with the same speaker until the other speaker takes initiative.

We classified a given turn as a question, prompt, or assertion/command where: a question tries to elicit information, a prompt doesn't express any propositional content, an assertion states facts, and a command intends to instigate action (Walker and Whittaker, 1990). We used the rules below to automatically annotate the turns:

- Question: if the turn ends with a question mark (?)

- Prompt: if a turn consists only of words 'uhhuh', 'okay', 'ok', 'yeah', 'umhmm', 'right', 'oh', 'umm', 'uh', 'hmm', 'umumm', 'ummm', 'alright'

- Assertion/Command: everything else

We didn't separately annotate command and assertion as we were more interested in the number of questions and prompts by the educators and patients; additionally, it would be hard to distinguish them using simple rules.

The rules for control transfer used by us are shown below (Turn type: Controller):

- Assertion/Command or Question: speaker unless response to a question

- prompt: hearer

Figure 3 shows the excerpt from Figure 1 marked with utterance type and control transfer. The utterances with type *assertion/command* indicate speaker initiative. On analyzing the transcripts, we found that on an average per conversation, educators asked 26 questions and produced 17 utterances with prompts as compared to 3 questions and 39 prompt utterances by the patients. As a consequence, an educator held the initiative for about 95 turns per conversation, whereas the patient did for 51 turns; the control lasted for about 5 turns on average in the case of an educator as compared to patients who only held control for 2 turns on average. These observations about patient/educator interactions have also been confirmed by an expert we have consulted with, Dr. Kishonna Gray from Department of Communication and Gender and Women's Studies at University of Illinois Chicago.

We hypothesize that, even if few, the turns where the patient takes control are important for the development of the dialogue agent: in fact, we envision this dialogue agent as an assistant that the patient will have to interact with on their initiative, rather than a system that operates as a health educator per se. Next, we will extract the topics from the turns where the patients have control since those are probable topics of discussion. We also plan to conduct focus groups with 10 self-identified AA and 10 H/L patients to gain insight into their lives post HF diagnosis and evaluate the acceptability of a dialogue agent to discuss HF. We believe talking to individuals with HF outside the hospital environment can help solicit questions that do not appear in the recordings or existing literature.

5   Conclusions and Future Work

In this paper, we discussed our data collection process for heart failure education conversations between educators and African American or Hispanic/Latino patients. We analyzed 20 transcribed recordings and found that the most common topic of discussion was food. Patients also discussed family and children frequently. Though mostly educators took the lead, we will extract topics where patients take control to build a dialogue agent that can answer patient queries effectively. We will also use insights from these interactions to inform the questions for the focus groups which we will conduct in the future.

Acknowledgments

We would like to thank Dr. Kishonna Gray for her valuable feedback and ideas on how to capture the cultural and inter-sectional complexities present with Hispanic/Latino and African American patients. This work is supported by the University of Illinois Chicago Discovery Partners Institute (DPI) Seed Funding Program.

References

Traci L. Alberti and Angela Nannini. 2013. Patient comprehension of discharge instructions from the emergency department: A literature review: Patient comprehension of discharge instructions from the ED. *Journal of the American Academy of Nurse Practitioners*, 25(4):186–194.

Matthew Barrett, Josiane Boyne, Julia Brandts, Hans-Peter Brunner-La Rocca, Lieven De Maesschalck, Kurt De Wit, Lana Dixon, Casper Eurlings, Donna Fitzsimons, Olga Golubnitschaja, Arjan Hageman, Frank Heemskerk, André Hintzen, Thomas M. Helms, Loreena Hill, Thom Hoedemakers, Nikolaus Marx, Kenneth McDonald, Marc Mertens, Dirk Müller-Wieland, Alexander Palant, Jens Piesk, Andrew Pomazanskyi, Jan Ramaekers, Peter Ruff, Katharina Schütt, Yash Shekhawat, Chantal F. Ski, David R. Thompson, Andrew Tsirkin, Kay van der Mierden, Chris Watson, and Bettina Zippel-Schultz. 2019. Artificial intelligence supported patient self-care in chronic heart failure: a paradigm shift from reactive to predictive, preventive and personalised care. *EPMA Journal*, 10(4):445–464.

Timothy W. Bickmore, Daniel Schulman, and Candace Sidner. 2013a. Automated interventions for multiple health behaviors using conversational agents. *Patient Education and Counseling*, 92(2):142–148.

Timothy W. Bickmore, Rebecca A. Silliman, Kerrie Nelson, Debbie M. Cheng, Michael Winter, Lori Henault, and Michael K. Paasche-Orlow. 2013b. A randomized controlled trial of an automated exercise coach for older adults. *Journal of the American Geriatrics Society*, 61(10):1676–1683.

David M Blei, Andrew Y Ng, and Michael I Jordan. 2003. Latent dirichlet allocation. *Journal of Machine Learning Research*, pages 993–1022.

Loretta Bobo, Robin J Womeodu, and Alfred L Knox Jr. 1991. Society of general internal medicine symposium principles of intercultural medicine in an internal medicine program. *The American Journal of the Medical Sciences*, 302(4):244–248.

Cheryl Der Ananian, Donna Winham, Sharon Thompson, and Megan Tisue. 2018. Perceptions of heart-healthy behaviors among african american adults: a mixed methods study. *International Journal of Environmental Research and Public Health*, 15(11):2433.

Ethan Fast, Binbin Chen, and Michael S Bernstein. 2016. Empath: Understanding topic signals in large-scale text. In *Proceedings of the 2016 CHI Conference on Human Factors in Computing Systems*, pages 4647–4657.

George Ferguson, James Allen, Lucian Galescu, Jill Quinn, and Mary Swift. 2009. CARDIAC: An Intelligent Conversational Assistant for Chronic Heart Failure Patient Heath Monitoring. In *2009 AAAI Fall Symposium Series*.

Teresa González-Lee and Harold J Simon. 1987. Teaching spanish and cross-cultural sensitivity to medical students. *Western Journal of Medicine*, 146(4):502.

Howie-Esquivel. 2014. A culturally appropriate educational intervention can improve self-care in hispanic patients with heart failure: a pilot randomized controlled trial. *Cardiology Research*.

Hannah Anderson Hughes and Bradi B. Granger. 2014. Racial disparities and the use of technology for self-management in blacks with heart failure: a literature review. *Current Heart Failure Reports*, 11(3):281–289.

Sharon Ann Hunt, William T. Abraham, Marshall H. Chin, Arthur M. Feldman, Gary S. Francis, Theodore G. Ganiats, Mariell Jessup, Marvin A. Konstam, Donna M. Mancini, Keith Michl, John A. Oates, Peter S. Rahko, Marc A. Silver, Lynne Warner Stevenson, and Clyde W. Yancy. 2009. 2009 focused update incorporated into the ACC/AHA 2005 guidelines for the diagnosis and management of heart failure in adults. *Journal of the American College of Cardiology*, 53(15).

Teresa D LaFromboise, Hardin LK Coleman, and Alexis Hernandez. 1991. Development and factor structure of the cross-cultural counseling inventory—revised. *Professional psychology: Research and practice*, 22(5):380.

Liliana Laranjo, Adam G Dunn, Huong Ly Tong, Ahmet Baki Kocaballi, Jessica Chen, Rabia Bashir, Didi Surian, Blanca Gallego, Farah Magrabi, Annie Y S Lau, and Enrico Coiera. 2018. Conversational agents in healthcare: a systematic review. *Journal of the American Medical Informatics Association*, 25(9):1248–1258.

Christopher S. Lee, Debra K. Moser, Terry A. Lennie, and Barbara Riegel. 2011. Event-free survival in adults with heart failure who engage in self-care management. *Heart & Lung*, 40(1):12–20.

Karina Lovell, Peter Bower, Judith Gellatly, Sarah Byford, Penny Bee, Dean McMillan, Catherine Arundel, Simon Gilbody, Lina Gega, Gillian Hardy, Shirley Reynolds, Michael Barkham, Patricia Mottram, Nicola Lidbetter, Rebecca Pedley, Jo Molle, Emily Peckham, Jasmin Knopp-Hoffer, Owen Price, Janice Connell, Margaret Heslin, Christopher Foley, Faye Plummer, and Christopher Roberts. 2017. Low-intensity cognitive-behaviour therapy interventions for obsessive-compulsive disorder compared to waiting list for therapist-led cognitive-behaviour therapy: 3-arm randomised controlled trial of clinical effectiveness. *PLOS Medicine*, 14(6):e1002337.

Sanjoy Moulik. 2019. *DIL - A Conversational Agent for Heart Failure Patients*. Ph.D. thesis.

Barbara Riegel, Beverly Carlson, Debra K Moser, Marge Sebern, Frank D Hicks, and Virginia Roland. 2004. Psychometric testing of the self-care of heart failure index. *Journal of Cardiac Failure*, 10(4):350–360.

Véronique L. Roger. 2013. Epidemiology of heart failure. *Circulation Research*, 113(6):646–659.

Gargi Roysircar Sodowsky, Richard C Taffe, Terry B Gutkin, and Steven L Wise. 1994. Development of the multicultural counseling inventory: A self-report measure of multicultural competencies. *Journal of Counseling Psychology*, 41(2):137.

SR Toukhsati, Tiny Jaarsma, AS Babu, Andrea Driscoll, and DL Hare. 2019. Self-care interventions that reduce hospital readmissions in patients with heart failure; towards the identification of change agents. *Clinical Medicine Insights: Cardiology*, 13:117954681985685.

Carolyn M. Tucker, Michael Marsiske, Kenneth G. Rice, Jessica Jones Nielson, and Keith Herman. 2011. Patient-centered culturally sensitive health care: Model testing and refinement. *Health Psychology*, 30(3):342–350.

Marilyn Walker and Steve Whittaker. 1990. Mixed initiative in dialogue: An investigation into discourse segmentation. In *Proceedings of the 28th annual meeting on Association for Computational Linguistics*, pages 70–78. Association for Computational Linguistics.

On the Utility of Audiovisual Dialog Technologies and Signal Analytics for Real-time Remote Monitoring of Depression Biomarkers

Michael Neumann[*]**, Oliver Roesler**[*]**, David Suendermann-Oeft**[*] **and Vikram Ramanarayanan**[*†]

[*] Modality.ai, Inc.

[†] University of California, San Francisco

`vikram.ramanarayanan@modality.ai`

Abstract

We investigate the utility of audiovisual dialog systems combined with speech and video analytics for real-time remote monitoring of depression at scale in uncontrolled environment settings. We collected audiovisual conversational data from participants who interacted with a cloud-based multimodal dialog system, and automatically extracted a large set of speech and vision metrics based on the rich existing literature of laboratory studies. We report on the efficacy of various audio and video metrics in differentiating people with mild, moderate and severe depression, and discuss the implications of these results for the deployment of such technologies in real-world neurological diagnosis and monitoring applications.

1 Introduction

Diagnosis, detection and monitoring of neurological and mental health in patients remain a critical need today. This necessitates the development of technologies that improve individuals' health and well-being by continuously monitoring their status, rapidly diagnosing medical conditions, recognizing pathological behaviors, and delivering just-in-time interventions, all in the user's natural information technology environment (Kumar et al., 2012). However, early detection or progress monitoring of neurological or mental health conditions, such as clinical depression, Amyotrophic Lateral Sclerosis (ALS), Alzheimer's disease, dementia, etc., is often challenging for patients due to multiple reasons, including, but not limited to: (i) lack of access to neurologists or psychiatrists; (ii) lack of awareness of a given condition and the need to see a specialist; (iii) lack of an effective standardized diagnostic or endpoint for many of these health conditions; (iv) substantial transportation and cost involved in conventional or traditional solutions; and in some cases, (v) shortage of medical specialists in these fields to begin with (Steven and Steinhubl, 2013).

We developed NEMSI (Suendermann-Oeft et al., 2019), or the NEurological and Mental health Screening Instrument, to bridge this gap. NEMSI is a cloud-based multimodal dialog system that conducts automated screening interviews over the phone or web browser to elicit evidence required for detection or progress monitoring of the aforementioned conditions, among others. While intelligent virtual agents have been proposed in earlier work for such diagnosis and monitoring purposes, NEMSI makes novel contributions along three significant directions: First, NEMSI makes use of devices available to everyone everywhere (web browser, mobile app, or regular phone), as opposed to dedicated, locally administered hardware, like cameras, servers, audio devices, etc. Second, NEMSI's backend is deployed in an automatically scalable cloud environment allowing it to serve an arbitrary number of end users at a small cost per interaction. Thirdly, the NEMSI system is natively equipped with real-time speech and video analytics modules that extract a variety of features of direct relevance to clinicians in the neurological and mental spaces.

A number of recent papers have investigated automated speech and machine vision features for predicting severity of depression (see for example France et al., 2000; Joshi et al., 2013; Meng et al., 2013; Jain et al., 2014; Kaya et al., 2014; Nasir et al., 2016; Pampouchidou al., 2016; Yang et al., 2017). These include speaking rate, duration, amplitude, and voice source/spectral features (fundamental frequency (F0), amplitude modulation, formants, and energy/power spectrum, among others) computed from the speech signal, and facial dynamics (for instance, landmark/facial action unit motions, global head motion, and eye blinks) and statistically derived features from emotions, action units, gaze, and pose derived from the video signal. We use these studies to inform our choices of speech and video metrics computed in real time,

Proceedings of the 1st Workshop on NLP for Medical Conversations, pages 47–52

Online, 10, 2020. ©2020 Association for Computational Linguistics

https://doi.org/10.18653/v1/P17

allowing clinicians to obtain useful analytics for their patients moments after they have interacted with the NEMSI dialog system.

We need to factor in additional considerations while deploying analytics modules as part of scalable real-time cloud-based systems in practice. Many of the studies above analyzed data recorded either offline or in laboratory conditions, implicitly assuming signal conditions which may hold differently or not at all during real world use. These considerations include, but are not limited to: (i) wide range of acoustic environments and lighting conditions resulting in variable background noise and choppy/blocky video at the user's end[1], (ii) limitations on a given user's network connection bandwidth and speed; (iii) the quantum of server traffic (or the number of patients/users trying to access the system simultaneously); and (iv) device calibration issues, given the wide range of user devices. This paper investigates the utility of a subset of audio and video biomarkers for depression collected using the NEMSI dialog system in such real-world conditions.

The rest of this paper is organized as follows: Sections 2 and 3 first present the NEMSI dialog system and the data collected and analyzed. Section 4 then details the speech and video feature extraction process. Section 5 presents statistical analyses of different groups of depression cohorts as determined by the reported PHQ-8 score, before Section 6 rounds out the paper, discussing the implications of our observations for real-world mental health monitoring systems.

2 System

2.1 NEMSI dialog ecosystem

NEMSI (NEurological and Mental health Screening Instrument) is a cloud-based multimodal dialog system. Refer to Suendermann-Oeft et al. (2019) for details regarding the system architecture and various software modules.

NEMSI end users are provided with a website link to the secure screening portal as well as login credentials by their caregiver or study liaison (physician or clinic). Once appropriate microphone and camera checks that the captured audio and video are of sufficient quality are complete, users hear the dialog agent's voice and are prompted to start a conversation with the agent, whose virtual

image also appears in a web window. Users are also able to see their own video, if so needed, in a small window in the upper right corner of the screen. The virtual agent then engages with users in a conversation using a mixture of structured speaking exercises and open-ended questions to elicit speech and facial behaviors relevant for the type of condition being screened for.

Analytics modules extract multiple speech (for instance, speaking rate, duration measures, F0, etc.) and video features (such as range and speed of movement of various facial landmarks) and store them in a database, along with information about the interaction itself such as the captured user responses, call duration, completion status, etc. All this information can be accessed by the clinicians after the interaction is completed through an easy-to-use dashboard which provides a high-level overview of the various aspects of the interaction (including the video thereof and analytic measures computed), as well as a detailed breakdown of the individual sessions and the underlying interaction turns.

3 Data

Depending on the health condition to be monitored and on the clinician's needs, different protocols can easily be employed in the NEMSI system. For the present study, we designed a protocol targeting the assessment of depression severity, based on (Mundt et al., 2007). The protocol elicits five different types of speech samples from participants that are consistently highlighted in the literature: (a) free speech (open-ended questions about subjects' emotional and physical state), (b) automated speech (counting up from 1), (c) read speech, (d) sustained vowels, and (e) measure of diadochokinetic rate (rapidly repeating the syllables /pa ta ka/).

After dialog completion, participants are asked to answer the Patient Health Questionnaire eight-item depression scale (PHQ-8), a standard scoring system for depression assessment (Kroenke et al., 2009). The self-reported PHQ-8 score serves as a reference point for our analysis. Further, we ask for information about age, sex, primary language and residence.

In total, we collected data from 307 interactions. After automatic data cleaning[2], 208 sessions re-

---

[1]Such conditions often arise despite explicit instructions to the contrary.

[2]We removed interactions for which PHQ-8 answers or relevant speech metrics were missing and sessions for which no face was detected in the video

mained for analysis. From those 208 participants, 98 were females, 97 were males and 13 did not specify. Mean participant age is 36.5 (SD = 12.1). 184 participants specified English as their primary language, 9 other languages and 15 did not specify. 176 participants were located in the US, 8 in the UK, 5 in Canada, 4 in other countries and 15 did not specify. Figure 1 shows the distribution of PHQ-8 scores among women and men.

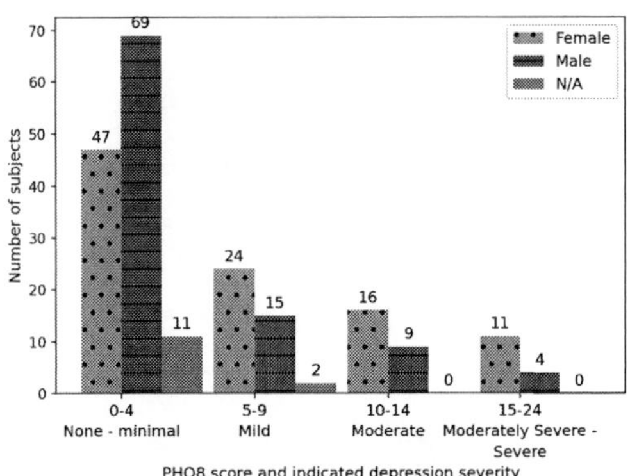

Figure 1: Distribution of PHQ-8 scores by gender.

Cutpoint (group sizes)	Free speech	Held Vowels
5 (127/81)	Percent pause time (a,f)	Volume (a,f), HNR (m), Mean F0 (m)
10 (168/40)	-	Jitter (f)
15 (193/15)	Volume (a,f,m)	Mean F0 (a), Volume (f)

Table 1: Speech metrics for which a statistically significant ($p < 0.05$) difference between sample populations is observed. In parentheses: f - females, m - males, a - all.

	Free speech	Read speech	Auto-mated	Held vowels	DDK
SpRate		✓			
ArtRate		✓			
SylRate					✓
PPT	✓	✓	✓		
Mean F0				✓	
Jitter				✓	
HNR				✓	
Volume	✓	✓	✓	✓	✓
Shimmer				✓	

Table 2: Speech metrics for each type of speech sample. SpRate = speaking rate, ArtRate = articulation rate, SylRate = syllable rate, PPT = percent pause time, DDK = dysdiadochokinesia.

4 Signal Processing and Metrics Extraction

4.1 Speech Metrics

For the speech analysis, we focus on *timing measures*, such as speaking rate and percentage of pause duration, *frequency domain measures*, such as fundamental frequency (F0) and jitter, and *energy-related measures*, such as volume and shimmer. We have selected commonly established speech metrics for clinical voice analysis (France et al., 2000; Mundt et al., 2012, 2007).

As described in Section 3, there are different types of speech samples, e.g. free speech and sustained vowels. Not all acoustic measures are meaningful for each type of stimuli. Table 2 presents all extracted metrics for the particular speech sample types.

All metrics are extracted with Praat (Boersma and Van Heuven, 2001). For the following measures, heuristics have been used to ignore obvious outliers in the analysis: articulation rate (excluded >350 words/min), speaking rate (excluded >250 words/min), percent pause time (excluded >80%).

4.2 Visual Metrics

For each utterance, 14 facial metrics were calculated in three steps: (i) face detection, (ii) facial landmark extraction, and (iii) facial metrics calculation. For face detection, the Dlib[3] face detector was employed, which uses 5 histograms of oriented gradients to determine the (x, y)-coordinates of one or more faces for every input frame (Dalal and Triggs, 2005). For facial landmark detection the Dlib facial landmark detector was employed, which uses an ensemble of regression trees proposed by Kazemi and Sullivan (2014), to extract 68 facial landmarks according to MultiPIE (Gross et al., 2010). Figure 2 illustrates the 14 facial landmarks: RB (right eyebrow), URER (right eye, upper right), RERC (right eye, right corner), LRER (right eye, lower right), LB (left eyebrow), ULEL (left eye, upper left), LELC (left eye, left corner), LLEL (left eye, lower left), NT (nose tip), UL (upper lip center), RC (right corner of mouth), LC (left corner of mouth), LL (lower lip center), and JC (jaw center). These are then used to calculate the following facial metrics:

---

[3] http://dlib.net/

49

Cutpoint	Gender	Free speech	Read speech
5	All	width, vJC, S_R, S_ratio, utter_dur	S_R
5	Female	S_ratio, utter_dur	eye_blinks
5	Male	S_ratio, eyebrow_vpos, eye_open, eye_blinks	
10	All	S_ratio, utter_dur	S_R
10	Female	open, width, LL_path, JC_path, S, S_R, S_L, eyebrow_vpos, eye_open	
10	Male	open, width, LL_path, JC_path, S, S_R, S_L, S_ratio, eyebrow_vpos, eye_open, eye_blinks	open, width, S, S_R, S_L, eyebrow_vpos
15	All	vLL, vJC	vLL, S_ratio
15	Female	width, vLL, vJC, S_ratio	vLL, S_ratio
15	Male	width, S, S_L, eyebrow_vpos, eye_open, eye_blinks	eyebrow_vpos

Table 3: Facial metrics for which a statistically significant ($p < 0.05$) difference between sample populations is observed. For gender *All* not only female and male samples, but also samples for which no gender was reported are used.

- **Movement measures**: Average lips opening and width (open, width) were calculated as the Euclidean distances between UL and LL, and RC and LC, respectively. Average displacement of LL and JC (LL_path, JC_path) were calculated as the module of the vector between the origin and LL and JC. Average eye opening (eye_open) was calculated as the Euclidean distances between URER and LRER, and ULEL and LLEL. Average vertical eyebrow displacement (eyebrow_vpos) was calculated as the difference between the vertical positions of RB and NT, and LB and NT. All measures were computed in millimeters.

- **Velocity measures**: The average velocity of LL and JC (vLL, vJC) in mm/s was calculated as the first derivative of LL_path and JC_path with time.

- **Surface measures**: The average total mouth surface (S) in mm^2 was calculated as the sum of the surfaces of the two triangles with vertices RC, UL, LL (S_R) and LC, UL, LL (S_L). Additionally, the mean symmetry ratio (S_ratio_avg) between S_R and S_L was determined.

- **Duration measures**: Utterance duration (utter_dur) in seconds.

- **Eye blink measures**: The number of eye blinks (eye_blinks) in blinks per second calculated using the eye aspect ratio as proposed by Soukupová and Čech (2016).

5 Analyses and Observations

The central research question of this study is the following: for a given metric, is there a statistically

Figure 2: Illustration of the 68 obtained and 14 used facial landmarks.

significant difference between participant cohorts with and without depression of a given severity (i.e., above and below a certain cut-point PHQ-8 score)? The PHQ-8 has established cutpoints above which the probability of a major depression increases substantially (Kroenke and Spitzer, 2002). Ten is commonly recommended as cutpoint for defining current depression (see (Kroenke et al., 2009) for a comprehensive overview). For our analysis, we use the cutpoints 5, 10, and 15 which align with the PHQ-8 score intervals of mild, moderate and moderately severe depression. Concretely, for each metric and cutpoint, we divide the data into two sample populations: (a) PHQ-8 score below and (b) PHQ-8 equal or above the cutpoint. We conducted a non-parametric Kruskal-Wallis test for every combination to find out whether certain obtained metrics show a statistically significant difference between cohorts.[4]

---

[4]We decided to exclude the /pa ta ka/ exercise (measure of diadochokineic rate) from the analysis, because we observed that many participants did not execute it correctly (e.g. making pauses between repetitions).

5.1 Analysis of Speech Metrics

Table 1 presents the acoustic measures and speech sample types, for which a significant difference between sample populations was observed ($p < 0.05$). For read speech, there is no significant difference for any of the metrics. For free speech, percentage of pause time and volume are indicators to distinguish groups. For sustained vowels, we observe significant differences for volume, mean fundamental frequency, harmonics-to-noise ratio and jitter. There are differences between females and males, as indicated in the table.

5.2 Analysis of Visual Metrics

Table 3 shows the visual metrics for which a significant difference between sample populations was observed ($p < 0.05$) for free and read speech. Visual metrics are only analyzed for free speech and read speech because only limited movement of facial muscles can be observed for automated speech and sustained vowels. For read speech only, a few metrics show significant differences independent of the cutpoint and gender, while the number of metrics for free speech depends on both cutpoint and gender. For males, the measures that involve the eyes, i.e. eye_open, eyebrow_vpos and eye_blinks, show significant differences independent of the employed cutpoint. In contrast, when considering all samples, independent of the reported gender, and females, the metrics for which significant differences are observed depend on the cutpoint and speech sample. Cutpoint 5 mostly includes eye, surface and duration measures, while cutpoint 10 also includes movement measures. For cutpoint 15, significant differences can be observed for the velocity of the lower lip and jaw center for both free and read speech, when considering all samples or females.

6 Conclusion and Outlook

We investigated whether various audio and video metrics extracted from audiovisual conversational data obtained through a cloud-based multimodal dialog system exhibit statistically significant differences between depressed and non-depressed populations. For several of the investigated metrics such differences were observed indicating that the employed audiovisual dialog system has a potential to be used for remote monitoring of depression. However, more detailed investigations on the nature of value distributions of metrics, their dependency on subject age or native language, the quality of input signals or used devices, among other studies, are necessary to to see to which degree the results are generalizable. Additionally, the used PHQ-scores were self-reported and might therefore be less accurate than scores obtained under the supervision of a clinician. In future work, we will also collect additional interactions from larger and more diverse populations. Furthermore, we will perform additional analysis on the obtained data, such as regression analysis. Finally, we will extend the set of investigated metrics and investigate their efficacy for other neurological or mental health conditions.

References

Paul Boersma and Vincent Van Heuven. 2001. Speak and unspeak with praat. *Glot International*, 5(9/10):341–347.

N. Dalal and B. Triggs. 2005. Histograms of oriented gradients for human detection. In *Proceedings of the IEEE Computer Society Conference on Computer Vision and Pattern Recognition (CVPR)*, San Diego, USA.

Daniel Joseph France, Richard G Shiavi, Stephen Silverman, Marilyn Silverman, and M Wilkes. 2000. Acoustical properties of speech as indicators of depression and suicidal risk. *IEEE transactions on Biomedical Engineering*, 47(7):829–837.

R. Gross, I. Matthews, J. Cohn, T. Kanade, and S. Baker. 2010. Multi-pie. In *Proceedings of the International Conference on Automatic Face and Gesture Recognition (FG)*, volume 28, pages 807–813.

Varun Jain, James L Crowley, Anind K Dey, and Augustin Lux. 2014. Depression estimation using audiovisual features and fisher vector encoding. In *Proceedings of the 4th International Workshop on Audio/Visual Emotion Challenge*, pages 87–91.

Jyoti Joshi, Roland Goecke, Sharifa Alghowinem, Abhinav Dhall, Michael Wagner, Julien Epps, Gordon Parker, and Michael Breakspear. 2013. Multimodal assistive technologies for depression diagnosis and monitoring. *Journal on Multimodal User Interfaces*, 7(3):217–228.

Heysem Kaya, Florian Eyben, Albert Ali Salah, and Björn Schuller. 2014. Cca based feature selection with application to continuous depression recognition from acoustic speech features. In *2014 IEEE International Conference on Acoustics, Speech and Signal Processing (ICASSP)*, pages 3729–3733. IEEE.

V. Kazemi and J. Sullivan. 2014. One millisecond face alignment with an ensemble of regression trees. In *IEEE Conference on Computer Vision and Pattern Recognition (CVPR)*, Columbus, USA.

Kurt Kroenke and Robert L Spitzer. 2002. The phq-9: a new depression diagnostic and severity measure. *Psychiatric annals*, 32(9):509–515.

Kurt Kroenke, Tara W Strine, Robert L Spitzer, Janet BW Williams, Joyce T Berry, and Ali H Mokdad. 2009. The phq-8 as a measure of current depression in the general population. *Journal of affective disorders*, 114(1-3):163–173.

Santosh Kumar, Wendy Nilsen, Misha Pavel, and Mani Srivastava. 2012. Mobile health: Revolutionizing healthcare through transdisciplinary research. *Computer*, 46(1):28–35.

Hongying Meng, Di Huang, Heng Wang, Hongyu Yang, Mohammed Ai-Shuraifi, and Yunhong Wang. 2013. Depression recognition based on dynamic facial and vocal expression features using partial least square regression. In *Proceedings of the 3rd ACM international workshop on Audio/visual emotion challenge*, pages 21–30.

James C Mundt, Peter J Snyder, Michael S Cannizzaro, Kara Chappie, and Dayna S Geralts. 2007. Voice acoustic measures of depression severity and treatment response collected via interactive voice response (ivr) technology. *Journal of neurolinguistics*, 20(1):50–64.

James C Mundt, Adam P Vogel, Douglas E Feltner, and William R Lenderking. 2012. Vocal acoustic biomarkers of depression severity and treatment response. *Biological psychiatry*, 72(7):580–587.

Md Nasir, Arindam Jati, Prashanth Gurunath Shivakumar, Sandeep Nallan Chakravarthula, and Panayiotis Georgiou. 2016. Multimodal and multiresolution depression detection from speech and facial landmark features. In *Proceedings of the 6th International Workshop on Audio/Visual Emotion Challenge*, pages 43–50.

Anastasia Pampouchidou, Olympia Simantiraki, Amir Fazlollahi, Matthew Pediaditis, Dimitris Manousos, Alexandros Roniotis, Georgios Giannakakis, Fabrice Meriaudeau, Panagiotis Simos, Kostas Marias, et al. 2016. Depression assessment by fusing high and low level features from audio, video, and text. In *Proceedings of the 6th International Workshop on Audio/Visual Emotion Challenge*, pages 27–34.

Tereza Soukupová and J. Čech. 2016. Real-time eye blink detection using facial landmarks. In *21stComputer Vision Winter Workshop*, Rimske Toplice, Slovenia.

R Steven and M Steinhubl. 2013. Can mobile health technologies transform health care. *JAMA*, 92037(1):1–2.

David Suendermann-Oeft, Amanda Robinson, Andrew Cornish, Doug Habberstad, David Pautler, Dirk Schnelle-Walka, Franziska Haller, Jackson Liscombe, Michael Neumann, Mike Merrill, et al. 2019. Nemsi: A multimodal dialog system for screening of neurological or mental conditions. In *Proceedings of the 19th ACM International Conference on Intelligent Virtual Agents*, pages 245–247.

Le Yang, Dongmei Jiang, Xiaohan Xia, Ercheng Pei, Meshia Cédric Oveneke, and Hichem Sahli. 2017. Multimodal measurement of depression using deep learning models. In *Proceedings of the 7th Annual Workshop on Audio/Visual Emotion Challenge*, pages 53–59.

Robust Prediction of Punctuation and Truecasing
for Medical ASR

Monica Sunkara   **Srikanth Ronanki**   **Kalpit Dixit**   **Sravan Bodapati**   **Katrin Kirchhoff**
Amazon AWS AI, USA
{sunkaral, ronanks}@amazon.com

Abstract

Automatic speech recognition (ASR) systems in the medical domain that focus on transcribing clinical dictations and doctor-patient conversations often pose many challenges due to the complexity of the domain. ASR output typically undergoes automatic punctuation to enable users to speak naturally, without having to vocalise awkward and explicit punctuation commands, such as "period", "add comma" or "exclamation point", while truecasing enhances user readability and improves the performance of downstream NLP tasks. This paper proposes a conditional joint modeling framework for prediction of punctuation and truecasing using pretrained masked language models such as BERT, BioBERT and RoBERTa. We also present techniques for domain and task specific adaptation by fine-tuning masked language models with medical domain data. Finally, we improve the robustness of the model against common errors made in ASR by performing data augmentation. Experiments performed on dictation and conversational style corpora show that our proposed model achieves ~5% absolute improvement on ground truth text and ~10% improvement on ASR outputs over baseline models under F1 metric.

1   Introduction

Medical ASR systems automatically transcribe medical speech found in a variety of use cases like physician-dictated notes (Edwards et al., 2017), telemedicine and even doctor-patient conversations (Chiu et al., 2017), without any human intervention. These systems ease the burden of long hours of administrative work and also promote better engagement with patients. However, the generated ASR outputs are typically devoid of punctuation and truecasing thereby making it difficult to comprehend. Furthermore, their recovery improves

the accuracy of subsequent natural language understanding algorithms (Peitz et al., 2011a; Makhoul et al., 2005) to identify information such as patient diagnosis, treatments, dosages, symptoms and signs. Typically, clinicians explicitly dictate the punctuation commands like "period", "add comma" etc., and a postprocessing component takes care of punctuation restoration. This process is usually error-prone as the clinicians may struggle with appropriate punctuation insertion during dictation. Moreover, doctor-patient conversations lack explicit vocalization of punctuation marks motivating the need for automatic prediction of punctuation and truecasing. In this work, we aim to solve the problem of automatic punctuation and truecasing restoration to medical ASR system text outputs.

Most recent approaches to punctuation and truecasing restoration problem rely on deep learning (Nguyen et al., 2019a; Salloum et al., 2017). Although it is a well explored problem in the literature, most of these improvements do not directly translate to great real world performance in all settings. For example, unlike general text, it is a harder problem to solve when applied to the medical domain for various reasons and we illustrate each of them:

- **Large vocabulary:** ASR systems in the medical domain have a large set of domain-specific vocabulary and several abbreviations. Owing to the domain specific data set and the open vocabulary in LVCSR (large-vocabulary continuous speech recognition) outputs, we often run into OOV (out of vocabulary) or rare word problems. Furthermore, a large vocabulary set leads to data sparsity issues. We address both these problems by using subword models. Subwords have been shown to work well in open-vocabulary speech recognition and several NLP tasks (Sennrich et al., 2015;

53

Proceedings of the 1st Workshop on NLP for Medical Conversations, pages 53–62
Online, 10, 2020. ©2020 Association for Computational Linguistics
https://doi.org/10.18653/v1/P17

Bodapati et al., 2019). We compare word and subword models across different architectures and show that subword models consistently outperform the former.

- **Data scarcity:** Data scarcity is one of the major bottlenecks in supervised learning. When it comes to the medical domain, obtaining data is not as straight-forward as some of the other domains where abundance of text is available. On the other hand, obtaining large amounts of data is a tedious and costly process; procuring and maintaining it could be a challenge owing to the strict privacy laws. We overcome the data scarcity problem, by using pretrained masked language models like BERT (Devlin et al., 2018) and its successors (Liu et al., 2019; Yang et al., 2019) which have successfully been shown to produce state-of-the-art results when finetuned for several downstream tasks like question answering and language inference. We approach the prediction task as a sequence labeling problem and jointly learn punctuation and truecasing. We show that finetuning a pretrained model with a very small medical dataset ($\sim$500k words) has $\sim$5% absolute performance improvement in terms of F1 compared to a model trained from scratch. We further boost the performance by first finetuning the masked language model to the medical speech domain and then to the downstream task.

- **ASR Robustness:** Models trained on ground truth data are not exposed to typical errors in speech recognition and perform poorly when evaluated on ASR outputs. Our objective is to make the punctuation prediction and truecasing more robust to speech recognition errors and establish a mechanism to test the performance of the model quantitatively. To address this issue, we propose a data augmentation based approach using n-best lists from ASR.

The contributions of this work are:

- A general post-processing framework for conditional joint labeling of punctuation and truecasing for medical ASR (clinical dictation and conversations).

- An analysis comparing different embeddings that are suitable for the medical domain. An in-depth analysis of the effectiveness of using pretrained masked language models like BERT and its successors to address the data scarcity problem.

- Techniques for effective domain and task adaptation using Masked Language Model (*MLM*) finetuning of BERT on medical domain data to boost the downstream task performance.

- Method for enhancing robustness of the models via data augmentation with n-best lists (from ASR output) to the ground truth during training to improve performance on ASR hypothesis at inference time.

The rest of this paper is organized as follows. Section 2 presents related work on punctuation and truecasing restoration. Section 3 introduces the model architecture used in this paper and describes various techniques for improving accuracy and robustness. The experimental evaluation and results are discussed in Section 4 and finally, Section 5 presents the conclusions.

2 Related work

Several researchers have proposed a number of methodologies such as the use of probabilistic machine learning models, neural network models, and the acoustic fusion approaches for punctuation prediction. We review related work in these areas below.

2.1 Earlier methods

In earlier efforts, punctuation prediction has been approached by using finite state or hidden Markov models (Gotoh and Renals, 2000; Christensen et al., 2001a). Several other approaches addressed it as a language modeling problem by predicting the most probable sequence of words with punctuation marks inserted (Stolcke et al., 1998; Beeferman et al., 1998; Gravano et al., 2009). Some others used conditional random fields (CRFs) (Lu and Ng, 2010; Ueffing et al., 2013) and maximum entropy using n-grams (Huang and Zweig, 2002). The rise of stronger machine learning techniques such as deep and/or recurrent neural networks replaced these conventional models.

2.2 Using acoustic information

Some methods used only acoustic information such as speech rate, intonation, pause duration etc.,

(Christensen et al., 2001b; Levy et al., 2012). While pauses influence in the prediction of Comma, intonation helps in disambiguation between punctuation marks like period and exclamation. Although this seemed to work, the most effective approach is to combine acoustic information with lexical information at word level using force-aligned duration (Klejch et al., 2017). In this work, we only considered lexical input and a pretrained lexical encoder for prediction of punctuation and truecasing. The use of pretrained acoustic encoder and fusion with lexical outputs are possible extensions in future work.

2.3 Neural approaches

Neural approaches for punctuation and truecasing can be classified into two broad categories: sequence labeling based models and MT-based seq2seq models. These approaches have proven to be quite effective in capturing the contextual information and achieved huge success. While some approaches considered only punctuation prediction, some others jointly modeled punctuation and truecasing.

One set of approaches treated punctuation as a machine translation problem and used phrase based statistical machine translation systems to output punctuated and true cased text (Peitz et al., 2011b; Cho et al., 2012; Driesen et al., 2014). Inspired by recent end-to-end approaches, (Yi and Tao, 2019) proposed the use of self-attention based transformer model to predict punctuation marks as output sequence for given word sequences. Most recently, (Nguyen et al., 2019b) proposed joint modeling of punctuation and truecasing by generating words with punctuation marks as part of the decoding. Although seq2seq based approaches have shown a strong performance, they are intensive, demanding and are not suitable for production deployment at large scale.

For sequence labeling problem, each word in the input is tagged with a punctuation. If there is no punctuation associated with a word, a blank label is used and is often referred as "no punc". (Cho et al., 2015) used a combination of neural networks and CRFs for joint prediction of punctuation and disfluencies. With growing popularity in deep recurrent neural networks, LSTMs and BLSTMs with attention mechanism were introduced for punctuation restoration (Tilk and Alumäe, 2015, 2016). Later, (Pahuja et al., 2017) proposed joint training of punc-

tuation and truecasing using BLSTM models. This work addressed joint learning as two correlated tasks, and predicted punctuation and truecasing as two independent outputs. Our proposed approach is similar to this work, but we rather condition truecasing prediction on punctuation output; this is discussed in detail in Section 3.

Punctuation and casing restoration for speech/ASR outputs in the medical domain has not been explored extensively. Recently, (Salloum et al., 2017) proposed a sequence labeling model using bi-directional RNNs with an attention mechanism and late fusion for punctuation restoration to clinical dictation. To our knowledge, there has not been any work on medical conversations, and we aim to bridge the gap here with latest advances in NLP with large-scale pretrained language models.

3 Modeling : Conditional Joint labeling of Punctuation + Casing

We propose a postprocessing framework for conditional and joint learning of punctuation and truecasing prediction. Consider an input utterance $x_{1:T} = \{x_1, x_2, ..., x_T\}$, of length T and consisting of words x_i. The first step in our modeling process involves punctuation prediction as a sequence tagging task. Once the model predicts a probability distribution over punctuation, this along with the input utterance is fed in as input for predicting the case of a word x_i. We consider the punctuation to be independent of casing and a conditional dependence of the truecase of a word on punctuation given the learned input representations. Our plausible reasoning follows from this example sentence – "She took dance classes. She had no natural grace or sense of rhythm.". The word after the period is capitalized, which implies that punctuation information can help in better prediction of casing. A pair of punctuation and truecasing is assigned per word:

$$\mathbf{Pr}(\mathbf{p_{1:T}}, \mathbf{c_{1:T}} | \mathbf{x_{1:T}}) = \\ \mathbf{Pr}(\mathbf{p_{1:T}} | \mathbf{x_{1:T}}) \mathbf{Pr}(\mathbf{c_{1:T}} | \mathbf{p_{1:T}}, \mathbf{x_{1:T}}) \quad (1)$$

where $c_i \in C$, a fixed set of casing labels {Lower_Case, Upper_Case, All_Caps, Mixed_Case}, and $p_i \in P$, a fixed set of punctuation labels {Comma, Period, Question_Mark, No_Punct}.

3.1 Pretrained lexical encoder

We propose to use a pretrained model like BERT, trained on a large text corpus, as a lexical encoder

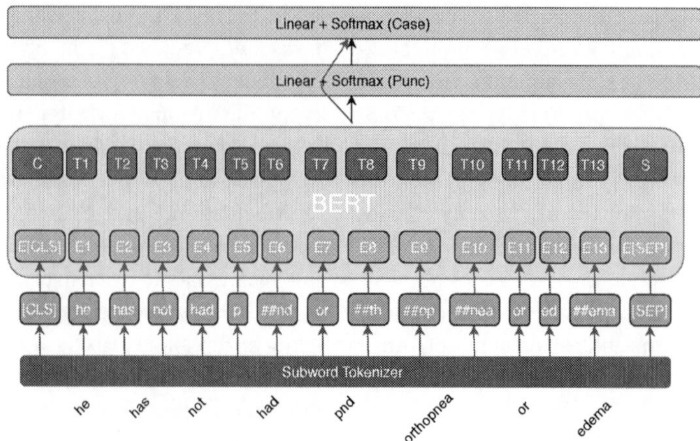

Figure 1: Pre-trained BERT encoder for prediction of punctuation and truecasing.

for learning an effective representation of the input utterance. Figure 1 illustrates our proposed model architecture.

Subword embeddings Given a sequence of input vectors $(x_1, x_2, ..., x_T)$, where x_i represents a word w_i, we extract the subword embeddings $(s_1, s_2, ..., s_n)$ using a wordpiece tokenizer (Schuster and Nakajima, 2012). Using subwords is especially effective in medical domain, as it contains more compound words with common subwords. For example consider the six words {hypotension, hypertension, hypoactive, hyperactive, active, tension } with four common subwords {hyper, hypo, active, tension}. In Section 4.2, we provide a comparative analysis of word and subword models across different architectures on medical data.

BERT encoder We provide subword embeddings $(s_1, s_2, ..., s_n)$ as input to the BERT encoder, which outputs a sequence of hidden states: $\mathbf{H} = (h1, ..., h_n)$ at its final layer. The pretrained BERT base encoder consists of 12 transformer encoder self-attention layers. For this task, we truncate the BERT encoder and fine-tune only the first six layers to reduce the model complexity. Although a deep encoder might enable us to learn a long memory context dependent representation of the input utterance, the performance gain is very minimal compared to the increased latency[1].

For punctuation, we input the last layer representations of truncated BERT encoder $h_1, h_2, ..., h_n$ to a linear layer with softmax activation to classify over the punctuation labels generating $(p_1, p_2, ..., p_n)$ as outputs. For casing, we concatenate the softmax probabilities of punctuation output with BERT encoder's outputs and feed to a linear layer with softmax activation generating case labels $(c_1, c_2, ..., c_n)$ for the sequence. The softmax output for punctuation $(\hat{p}_i)$ and truecasing $(\hat{c}_i)$ is as follows:

$$\hat{p}_i = softmax(W^k h_i + b^k) \qquad (2)$$

$$\hat{c}_i = softmax(W^l(\hat{p}_i \oplus h_i) + b^l) \qquad (3)$$

where W^k, b^k denote weights and bias of punctuation linear output layer and W^l, b^l denote weights and bias of truecasing linear output layer.

Joint learning objective: We model our learning objective to maximize the joint probability $\mathbf{Pr}(\mathbf{p_{1:T}}, \mathbf{c_{1:T}}|\mathbf{x_{1:T}})$. The model is finetuned end-to-end to minimize the cross-entropy loss between the assigned distribution and the training data. The parameters of BERT encoder are shared across punctuation and casing prediction tasks and are jointly trained. We compute the losses (L^p, L^c) for each task using cross entropy loss function. The final loss L to be optimized is a weighted average of the task-specific loses:

$$L = \alpha L^p + L^c \qquad (4)$$

where α is a fixed weight optimized for best predictions across both the tasks. In our experiments, we explored α values in the range of (0.2-2) and found 0.6 to be the optimal value.

---

[1]We experimentally found that 12-layer BERT base model gives ∼1% improvement over 6-layer BERT base model whereas the inference and training times were double for the former.

3.2 Finetuning using Masked Language Model with Medical domain data

BERT and its successors have shown great performance on downstream NLP tasks. But just like any other model, these Language Models are biased by their training data. In particular, they are typically trained on data that is easily available in large quantities on the internet e.g. Wikipedia, CommonCrawl etc. Our domain, Medical ASR Text, is not "common" and is very under-represented in the training data for these Language Models. One way to correct this situation is to perform a few steps of unsupervised Masked Language Model finetuning on the BERT models before performing cross-entropy training using the labeled task data (Han and Eisenstein, 2019).

Domain adaptation We finetune the pretrained BERT model for MLM (Masked LM) objective on medical domain data. 15% of input tokens are masked randomly before feeding into the BERT model as proposed by (Devlin et al., 2018). The main goal is to adapt and learn better representations of speech data. The domain adapted model can be further finetuned with an additional layer to a downstream task like punctuation and casing prediction.

Domain+Task adaptation Building on the previous technique, we attempt to finetune the pretrained model for task adaptation in combination with domain adaptation. In this technique, instead of randomly masking 15% of the input tokens, we do selective masking i.e. 50% of the masked tokens would be random and the other 50% would be punctuation marks ([".", ",", "?"] in our case). Therefore, the finetuned model would not only adapt to speech domain, but would also effectively learn the placement of punctuation marks in a text based on the context.

3.3 Robustness to ASR errors

Models trained on ground truth text inputs may not perform well when tested with ASR output, especially when the system introduces grammatical errors. To make models more robust against ASR errors, we perform data augmentation with ASR outputs for training. For punctuation restoration, we use edit distance measure to align ASR hypothesis with ground truth punctuated text. Before computing alignment, we strip all punctuation from ground truth and lowercase the text. This helps us find the best alignment between ASR hypothesis and ground truth text. Once the alignment is found, we restore the punctuation from each word in ground truth text to hypothesis. If there are words that are punctuated in ground truth but got deleted in ASR hypothesis, we restore the punctuation to previous word. For truecasing, we try to match the reference word with hypothesis word from aligned sequences with a window size of 5, two words to the left and two words to the right of current word and restore truecasing only in the cases where reference word is found. We performed experiments with data augmentation using 1-best hypothesis and n-best lists as additional training data and the results are reported in Section 4.4.

4 Experiments and results

4.1 Data

We evaluate our proposed framework and models on a subset of two internal medical datasets: dictation and conversational. The dictation corpus contains 3.7M words and the conversational corpus contains 51M words. The medical data comes with special tags masking personal identifiable and patient health information. We also use a general domain Wikipedia dataset for comparative analysis with Medical domain data. This data is a subset of the publicly available release of Wiki dataset (Sproat and Jaitly, 2016). The corpus contains 35M words and relatively shorter sentences ranging from 8 to 200 words in length. 90% of the data from each corpus is used for training, 5% for fine-tuning and remaining 5% is held-out for testing.

For robustness experiments presented in Section 4.4, we used data from the dictation corpus consisting of 2265 text files and corresponding audio files with an average duration of $\sim$15 minutes. The total length of the corpus is 550 hours. For augmentation with ground-truth transcription, we transcribed audio files using a speech recognition system. Restoration of punctuation and truecasing to transcribed text can be erroneous as the word error rate(WER) goes up. We therefore discarded the transcribed text of those audio files whose WER is more than 25%. We sorted the remaining transcriptions based on WER to make further splits: hypothesis from top 50 files with best WER is set as test data, and the next 50 files were chosen as development and rest of the transcribed text was used for training. The partition was done this way to minimize the number of errors that may occur

		Punctuation			Truecasing			
Model	Token	No Punc	Full stop	Comma	LC	UC	CA	MC
CNN-Highway	word	0.97	0.81	0.71	0.98	0.84	0.95	0.99
	subword	0.98	0.83	0.70	0.99	0.87	0.95	0.99
3-LSTM	word	0.97	0.82	0.73	0.98	0.84	0.96	0.98
	subword	0.98	0.84	0.75	0.99	0.87	0.97	0.99
3-BLSTM	word	0.98	0.86	0.75	0.99	0.88	0.97	0.98
	subword	0.99	0.87	0.76	0.99	0.90	0.97	1.0
Transformer encoder	word	0.97	0.84	0.7	0.98	0.86	0.97	0.98
	subword	0.98	0.85	0.72	0.99	0.87	0.97	0.99

Table 1: Dictation corpus: Comparison of F1 scores for punctuation and truecasing across different model architectures using word and subword tokens (LC: lower case; UC: Upper case; CA: CAPS All; MC: Mixed Case).

		Punctuation				Truecasing			
Model	Token	No Punc	Full stop	Comma	QM	LC	UC	CA	MC
CNN-Highway	word	0.96	0.72	0.64	0.60	0.96	0.78	0.99	0.91
	subword	0.97	0.74	0.65	0.61	0.97	0.80	0.98	0.99
3-LSTM	word	0.96	0.74	0.64	0.65	0.96	0.79	0.99	0.95
	subword	0.97	0.75	0.65	0.66	0.97	0.79	0.97	1.0
3-BLSTM	word	0.97	0.77	0.68	0.68	0.97	0.82	0.99	0.95
	subword	0.98	0.79	0.68	0.69	0.97	0.83	0.99	1.0
Transformer encoder	word	0.97	0.77	0.68	0.68	0.97	0.83	0.99	0.92
	subword	0.98	0.79	0.69	0.69	0.98	0.83	0.99	1.0

Table 2: Conversational corpus: Comparison of F1 scores for punctuation and truecasing across different model architectures using word and subword tokens (QM: Question Mark; LC: lower case; UC: Upper case; CA: CAPS All; MC: Mixed Case).

during restoration.

Preprocessing long-speech transcriptions Conversational style speech has long-speech transcripts, in which the context is spread across multiple segments. we use an overlapped chunking and merging component to pre and post process the data. We use a sliding window approach (Nguyen et al., 2019a) to split long ASR outputs into chunks of 200 words each with an overlapping window of 50 words each to the left and right. The overlap helps in preserving the context for all the words after splitting and ensures accurate prediction of punctuation and case corresponding to each word.

4.2 Large Vocabulary: Word vs Subword models

For a fair comparison with BERT, we evaluate various recurrent and non-recurrent architectures with both word and subword embeddings. The two recurrent models include a 3 layer uni-directional LSTM (*3-LSTM*) and a 3 layer Bi-directional LSTM (*3-BLSTM*). One of the non recurrent encoders, implements a CNN-Highway architecture based on the work proposed by (Kim et al., 2016),

whereas the other one implements a transformer encoder based model (Vaswani et al., 2017). We train all four models on medical data from dictation and conversation corpus with weights initialized randomly. The vocabulary for word models is derived by considering all the unique words from training corpus, with additional tokens for unknown and padding. This yielded a vocabulary size of 30k for dictation and 64k for conversational corpus. Subwords are extracted using a wordpiece model (Schuster and Nakajima, 2012) and its inventory is less than half that of word model for conversation. Tables 1 and 2 summarize our results on dictation and conversation datasets respectively. We observe that subword models consistently performed same or better than word models. On punctuation task, for Full stop and Comma, we notice an absolute ~1-2% improvement respectively on dictation set. Similarly, on the conversation dataset, we notice an absolute ~1-2% improvement on Full stop, Comma and Question Mark. For the casing task, we notice that word and subword models performed equally well except in dictation dataset where we see an absolute ~3% improvement for Upper_Case.

Model	Dataset	Punctuation			Truecasing			
		No Punc	Full stop	Comma	LC	UC	CA	MC
3-BLSTM	Wiki	0.95	0.17	0.27	0.95	0.31	0.55	0.19
BERT	Wiki	0.96	0.2	0.39	0.95	0.36	0.65	0.2
3-BLSTM	Medical	0.99	0.87	0.76	0.99	0.9	0.97	1.0
BERT	Medical	0.99	0.9	0.81	0.99	0.93	0.99	1.0
FT-BERT	Medical	0.99	0.92	0.82	0.99	0.93	0.99	1.0
PM-BERT	Medical	0.99	0.93	0.82	0.99	0.94	0.99	1.0
Bio-BERT	Medical	0.99	0.92	0.82	0.99	0.93	0.99	1.0
RoBERTa	Medical	0.99	0.92	0.81	0.99	0.94	0.99	1.0

Table 3: Comparison of F1 scores for punctuation and truecasing using BERT and BLSTM when trained on Wiki data and Medical dictation data (FT-BERT: Finetuned BERT for domain adapation, PM-BERT: Finetuned BERT by punctuation masking for domain and task adapation).

Model	Dataset	Punctuation				Truecasing			
		No Punc	Full stop	Comma	QM	LC	UC	CA	MC
3-BLSTM	Wiki	0.89	0.001	0.25	0.002	0.93	0.13	0.9	0.95
BERT	Wiki	0.93	0.004	0.4	0.007	0.93	0.4	0.95	0.95
3-BLSTM	Medical	0.98	0.79	0.68	0.69	0.97	0.83	0.99	1.0
BERT	Medical	0.98	0.8	0.71	0.72	0.98	0.85	0.99	1.0
FT-BERT	Medical	0.98	0.81	0.72	0.73	0.98	0.85	0.99	1.0
PM-BERT	Medical	0.98	0.82	0.72	0.74	0.98	0.86	0.99	1.0
Bio-BERT	Medical	0.98	0.81	0.71	0.72	0.98	0.85	0.99	1.0
RoBERTa	Medical	0.98	0.82	0.73	0.74	0.98	0.86	0.99	1.0

Table 4: Comparison of F1 scores for punctuation and truecasing using BERT and BLSTM when trained on Wiki data and Medical conversation data (FT-BERT: Finetuned BERT for domain adapation, PM-BERT: Finetuned BERT by punctuation masking for domain and task adapation).

We hypothesize that medical vocabulary contains a large set of compound words, which a subword based model works effectively over word model. Upon examining few utterances, we noticed that subword models can learn effective representations of these compound medical words by tokenizing them into subwords. On the other hand, word models often run into rare word or OOV issues.

4.3 Pretrained language models

Significance of in-domain data For analyzing the importance of in-domain data, we train a baseline BLSTM model and a pretrained BERT model on Wiki and Medical data from both dictation and conversational corpus and tested the models on Medical held-out data. The first four rows of Tables 3 and 4 summarize the results. The models trained on Wiki data performed very poorly when compared to models trained on Medical data from either dictation or conversation corpus. Although dictation corpus (3.7M words) is relatively smaller than Wiki corpus (35M words), the difference in accuracy is significantly higher across both

models. Imbalanced classes like Full stop, Comma, Question_Mark were most affected. Another interesting observation is that the models trained on Medical data performed better on Full stop compared to Comma; whereas general domain models performed better on Comma compared to Full stop. The degradation in general models might be due to Wiki sentences being short and ending with a Full stop unlike lengthy medical transcripts. Also, the F1 scores are lower on conversation data across both the tasks, indicating the complexity involved in modeling conversational data due to their highly unstructured format. Overall, the pretrained BERT model consistently outperformed baseline BLSTM model on both dictation and conversation data. This motivated us to focus on adapting the pretrained models for this task.

Finetuning Masked LM We have run two levels of fine-tuning as explained in Section 3.2. First, we finetuned BERT with Medical domain data using random masking (*FT-BERT*) and for task adaptation, we performed fine-tuning with

Model	n-best	Punctuation				Truecasing			
		No Punc	Full stop	Comma	QM	LC	UC	CA	MC
BERT-GT	-	0.97	0.58	0.45	0.0	0.98	0.60	0.78	0.90
BERT-ASR	1-best	0.97	0.66	0.56	0.54	0.99	0.72	0.86	1.0
	3-best	0.98	0.67	0.57	0.42	0.98	0.69	0.79	0.84
	5-best	0.97	0.61	0.5	0.35	0.98	0.65	0.79	0.83

Table 5: Comparison of F1 scores for punctuation and truecasing with ground truth and ASR augmented data.

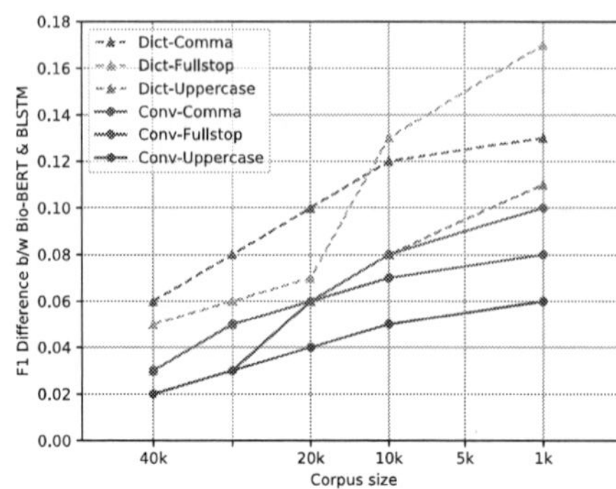

Figure 2: Difference in F1 scores between Bio-BERT and BLSTM for varying data sizes.

punctuation based masking (*PM-BERT*). For both experiments, we used the same data as we have used for finetuning the downstream task. From the results presented in Table 3 and 4, we infer that finetuning boosts the performance of punctuation and truecasing (an absolute improvement of $\sim$1-2%). From both the datasets, it is clear that task specific masking helps better than simple random masking. For dictation dataset, Full stop improved by an absolute 3% by performing punctuation specific masking, suggesting that finetuning MLM can give higher benefits when the amount of data is low.

Variants of BERT We compare three pretrained models namely, BERT and its successor RoBERTa (Liu et al., 2019) and Bio-BERT (Lee et al., 2020) which was trained on large scale Biomedical corpora. The results are summarized in last two rows of Table 3 and 4. First, we observe that both Bio-BERT and RoBERTa outperformed the initial BERT model and has shown an absolute $\sim$3-5% improvement over the baseline 3-BSLTM. To further validate this, we extended our experiments to understand how the performance of our best model(Bio-BERT) varies across different training dataset sizes compared to the baseline. From Figure 2, we observe that the difference increases significantly as we move towards smaller datasets. For the smallest data set size of 500k words (1k transcripts), there

is an absolute improvement of 6-17% over the baseline in accuracy in terms of F1. This shows that pretraining on a large dataset helps to overcome data scarcity issue effectively.

4.4 Robustness

For testing robustness, we performed experiments with augmentation of ASR data from n-best lists (*BERT-ASR*). We considered top-1, top-3 and top-5 hypotheses for n-best lists augmentation with ground truth text and the results are presented in Table 5. Additionally, the best BERT model trained using only ground truth text inputs (*BERT-GT*) from Table 3 is also evaluated on ASR outputs. To compute F1 scores on held-out test set, we first aligned the ASR hypothesis with ground truth data and restored the punctuation and truecasing as described in Section 3.3. From the results presented in Table 5, we infer that adding ASR hypothesis to the training data helped improve the performance of both punctuation and truecasing. In punctuation, both Full stop and Comma have seen an absolute 10% improvement in F1 score. Although the number of question marks is less in test data, the augmented systems performed really well compared to the system trained purely on ground truth text. However, we found that using n-best lists with $n > 1$ did not help much compared to the 1-best list. This may be due to sub-optimal restoration of punctuation and truecasing as the WER with n-best lists is likely to go up as n increases.

5 Conclusion

In this paper, we have presented a framework for conditional joint modeling of punctuation and truecasing in medical transcriptions using pretrained language models such as BERT. We also demonstrated the benefit from MLM objective finetuning of the pretrained model with task specific masking. We further improved the robustness of punctuation and truecasing on ASR outputs by data augmentation during training. Experiments performed on both dictation and conversation corpora show the effectiveness of the proposed approach. Future work includes the use of either pretrained acoustic features or pretrained acoustic encoder to perform fusion with pretrained linguistic encoder to further boost the performance of punctuation.

References

Doug Beeferman, Adam Berger, and John Lafferty. 1998. Cyberpunc: A lightweight punctuation annotation system for speech. In *Proceedings of the 1998 IEEE International Conference on Acoustics, Speech and Signal Processing, ICASSP'98 (Cat. No. 98CH36181)*, volume 2, pages 689–692. IEEE.

Sravan Bodapati, Spandana Gella, Kasturi Bhattacharjee, and Yaser Al-Onaizan. 2019. Neural word decomposition models for abusive language detection. In *Proceedings of the Third Workshop on Abusive Language Online*, pages 135–145, Florence, Italy. Association for Computational Linguistics.

Chung-Cheng Chiu, Anshuman Tripathi, Katherine Chou, Chris Co, Navdeep Jaitly, Diana Jaunzeikare, Anjuli Kannan, Patrick Nguyen, Hasim Sak, Ananth Sankar, et al. 2017. Speech recognition for medical conversations. *arXiv preprint arXiv:1711.07274*.

Eunah Cho, Kevin Kilgour, Jan Niehues, and Alex Waibel. 2015. Combination of nn and crf models for joint detection of punctuation and disfluencies. In *Sixteenth annual conference of the international speech communication association*.

Eunah Cho, Jan Niehues, and Alex Waibel. 2012. Segmentation and punctuation prediction in speech language translation using a monolingual translation system. In *International Workshop on Spoken Language Translation (IWSLT) 2012*.

Heidi Christensen, Yoshihiko Gotoh, and Steve Renals. 2001a. Punctuation annotation using statistical prosody models. In *ISCA tutorial and research workshop (ITRW) on prosody in speech recognition and understanding*.

Heidi Christensen, Yoshihiko Gotoh, and Steve Renals. 2001b. Punctuation annotation using statistical prosody models. In *ISCA tutorial and research workshop (ITRW) on prosody in speech recognition and understanding*.

Jacob Devlin, Ming-Wei Chang, Kenton Lee, and Kristina Toutanova. 2018. Bert: Pre-training of deep bidirectional transformers for language understanding. *arXiv preprint arXiv:1810.04805*.

Joris Driesen, Alexandra Birch, Simon Grimsey, Saeid Safarfashandi, Juliet Gauthier, Matt Simpson, and Steve Renals. 2014. Automated production of truecased punctuated subtitles for weather and news broadcasts. In *Fifteenth Annual Conference of the International Speech Communication Association*.

Erik Edwards, Wael Salloum, Greg P Finley, James Fone, Greg Cardiff, Mark Miller, and David Suendermann-Oeft. 2017. Medical speech recognition: reaching parity with humans. In *International Conference on Speech and Computer*, pages 512–524. Springer.

Yoshihiko Gotoh and Steve Renals. 2000. Sentence boundary detection in broadcast speech transcripts.

Agustin Gravano, Martin Jansche, and Michiel Bacchiani. 2009. Restoring punctuation and capitalization in transcribed speech. In *2009 IEEE International Conference on Acoustics, Speech and Signal Processing*, pages 4741–4744. IEEE.

Xiaochuang Han and Jacob Eisenstein. 2019. Unsupervised domain adaptation of contextualized embeddings for sequence labeling. In *Proceedings of the 2019 Conference on Empirical Methods in Natural Language Processing and the 9th International Joint Conference on Natural Language Processing (EMNLP-IJCNLP)*, pages 4229–4239.

Jing Huang and Geoffrey Zweig. 2002. Maximum entropy model for punctuation annotation from speech. In *Seventh International Conference on Spoken Language Processing*.

Yoon Kim, Yacine Jernite, David Sontag, and Alexander M Rush. 2016. Character-aware neural language models. In *AAAI*, pages 2741–2749.

Ondřej Klejch, Peter Bell, and Steve Renals. 2017. Sequence-to-sequence models for punctuated transcription combining lexical and acoustic features. In *2017 IEEE International Conference on Acoustics, Speech and Signal Processing (ICASSP)*, pages 5700–5704. IEEE.

Jinhyuk Lee, Wonjin Yoon, Sungdong Kim, Donghyeon Kim, Sunkyu Kim, Chan Ho So, and Jaewoo Kang. 2020. Biobert: a pre-trained biomedical language representation model for biomedical text mining. *Bioinformatics*, 36(4):1234–1240.

Tal Levy, Vered Silber-Varod, and Ami Moyal. 2012. The effect of pitch, intensity and pause duration in punctuation detection. In *2012 IEEE 27th Convention of Electrical and Electronics Engineers in Israel*, pages 1–4. IEEE.

Yinhan Liu, Myle Ott, Naman Goyal, Jingfei Du, Mandar Joshi, Danqi Chen, Omer Levy, Mike Lewis, Luke Zettlemoyer, and Veselin Stoyanov. 2019. Roberta: A robustly optimized bert pretraining approach. *arXiv preprint arXiv:1907.11692*.

Wei Lu and Hwee Tou Ng. 2010. Better punctuation prediction with dynamic conditional random fields. In *Proceedings of the 2010 conference on empirical methods in natural language processing*, pages 177–186.

John Makhoul, Alex Baron, Ivan Bulyko, Long Nguyen, Lance Ramshaw, David Stallard, Richard Schwartz, and Bing Xiang. 2005. The effects of speech recognition and punctuation on information extraction performance. In *Ninth European Conference on Speech Communication and Technology*.

Binh Nguyen, Vu Bao Hung Nguyen, Hien Nguyen, Pham Ngoc Phuong, The-Loc Nguyen, Quoc Truong Do, and Luong Chi Mai. 2019a. Fast and accurate capitalization and punctuation for automatic speech recognition using transformer and chunk merging. *arXiv preprint arXiv:1908.02404*.

Binh Nguyen, Vu Bao Hung Nguyen, Hien Nguyen, Pham Ngoc Phuong, The-Loc Nguyen, Quoc Truong Do, and Luong Chi Mai. 2019b. Fast and accurate capitalization and punctuation for automatic speech recognition using transformer and chunk merging. *arXiv preprint arXiv:1908.02404*.

Vardaan Pahuja, Anirban Laha, Shachar Mirkin, Vikas Raykar, Lili Kotlerman, and Guy Lev. 2017. Joint learning of correlated sequence labelling tasks using bidirectional recurrent neural networks. *arXiv preprint arXiv:1703.04650*.

Stephan Peitz, Markus Freitag, Arne Mauser, and Hermann Ney. 2011a. Modeling punctuation prediction as machine translation. In *International Workshop on Spoken Language Translation (IWSLT) 2011*.

Stephan Peitz, Markus Freitag, Arne Mauser, and Hermann Ney. 2011b. Modeling punctuation prediction as machine translation. In *International Workshop on Spoken Language Translation (IWSLT) 2011*.

Wael Salloum, Gregory Finley, Erik Edwards, Mark Miller, and David Suendermann-Oeft. 2017. Deep learning for punctuation restoration in medical reports. In *BioNLP 2017*, pages 159–164.

Mike Schuster and Kaisuke Nakajima. 2012. Japanese and korean voice search. In *2012 IEEE International Conference on Acoustics, Speech and Signal Processing (ICASSP)*, pages 5149–5152. IEEE.

Rico Sennrich, Barry Haddow, and Alexandra Birch. 2015. Neural machine translation of rare words with subword units. *arXiv preprint arXiv:1508.07909*.

Richard Sproat and Navdeep Jaitly. 2016. Rnn approaches to text normalization: A challenge. *arXiv preprint arXiv:1611.00068*.

Andreas Stolcke, Elizabeth Shriberg, Rebecca Bates, Mari Ostendorf, Dilek Hakkani, Madelaine Plauche, Gokhan Tur, and Yu Lu. 1998. Automatic detection of sentence boundaries and disfluencies based on recognized words. In *Fifth International Conference on Spoken Language Processing*.

Ottokar Tilk and Tanel Alumäe. 2015. Lstm for punctuation restoration in speech transcripts. In *Sixteenth annual conference of the international speech communication association*.

Ottokar Tilk and Tanel Alumäe. 2016. Bidirectional recurrent neural network with attention mechanism for punctuation restoration. In *Interspeech*, pages 3047–3051.

Nicola Ueffing, Maximilian Bisani, and Paul Vozila. 2013. Improved models for automatic punctuation prediction for spoken and written text. In *Interspeech*, pages 3097–3101.

Ashish Vaswani, Noam Shazeer, Niki Parmar, Jakob Uszkoreit, Llion Jones, Aidan N Gomez, Łukasz Kaiser, and Illia Polosukhin. 2017. Attention is all you need. In *Advances in neural information processing systems*, pages 5998–6008.

Zhilin Yang, Zihang Dai, Yiming Yang, Jaime Carbonell, Russ R Salakhutdinov, and Quoc V Le. 2019. Xlnet: Generalized autoregressive pretraining for language understanding. In *Advances in neural information processing systems*, pages 5754–5764.

Jiangyan Yi and Jianhua Tao. 2019. Self-attention based model for punctuation prediction using word and speech embeddings. In *Proc. ICASSP*, pages 7270–7274.

Topic-Based Measures of Conversation for Detecting Mild Cognitive Impairment

Liu Chen
Center for Spoken Language Understanding
Oregon Health & Science University
chliu@ohsu.edu

Hiroko H Dodge
Department of Neurology
Oregon Health & Science University
dodgeh@ohsu.edu

Meysam Asgari
Center for Spoken Language Understanding
Oregon Health & Science University
asgari@ohsu.edu

Abstract

Conversation is a complex cognitive task that engages multiple aspects of cognitive functions to remember the discussed topics, monitor the semantic and linguistic elements, and recognize others' emotions. In this paper, we propose a computational method based on the lexical coherence of consecutive utterances to quantify topical variations in semi-structured conversations of older adults with cognitive impairments. Extracting the lexical knowledge of conversational utterances, our method generates a set of novel conversational measures that indicate underlying cognitive deficits among subjects with mild cognitive impairment (MCI). Our preliminary results verify the utility of the proposed conversation-based measures in distinguishing MCI from healthy controls.

1 Introduction

Speech and language characteristics are known to be effective social behavioral markers that could potentially serve to facilitate the identification of measured "markers" reflecting early cognitive changes in at-risk older adults. Recent advances on natural language processing (NLP) algorithms have given the researchers the opportunity to explore subtleties of spoken language samples and extract a wider range of clinically useful measures. Leveraging an NLP-based method, our objective in this study is to characterize the ongoing dynamics of topics over the course of everyday conversation between an interviewer and an older adult with or without cognitive impairment. Our proposed method translates its analysis of conversation into a set of quantifiable measures that can be used in clinical trials for early detection of a cognitive deficit. Our cohort includes a professionally transcribed dataset of 30-minute audio recordings collected from conversation-based social interactions carried out between standardized interviewers and participants with either normal cognition or MCI (clinicaltirals.gov: NCT02871921). We evaluate the utility of proposed conversation-based measures in detecting MCI incidence. To the best of our knowledge, analysis of exchanged topics in conversations have not been used to examine the cognitive status of older adults.

1.1 Conversational Speech and Cognitive Impairment

Recent studies have attempted to leverage natural language processing (NLP) algorithms to automatically characterize atypical language characteristics observed in age-related cognitive decline(Roark et al., 2011; Asgari et al., 2017; Shibata et al., 2016; Mueller et al., 2016). With a few exceptions, most of these studies have used elicited speech paradigms to generate speech samples, for example, using traditional neuropsychological language tests such as the verbal fluency test (citing names from a semantic category such as animals or fruits within a short amount of time) or the story recall test (recalling specific stories subjects are exposed to during a testing session). As a result, their assessment of language characteristics is constrained by the nature of language tests. Alternatively, everyday conversations have been recently explored to gain insight about the consequences of a cognitive deficit on a patient's speech and language characteristics (Khodabakhsh et al., 2015; López-de Ipina et al., 2015; Hoffmann et al., 2010). Semi-structured conversations (i.e., talk about pre-specified topics) more closely resemble to naturalistic speech than elicited speech tasks (e.g., verbal fluency tests, picture naming tests) and provide a rich source of information allowing us to correlate various aspects of spoken language to cognitive functioning. Conversation is a complex cognitive task that engages multiple domains of

Proceedings of the 1st Workshop on NLP for Medical Conversations, pages 63–67
Online, 10, 2020. ©2020 Association for Computational Linguistics
https://doi.org/10.18653/v1/P17

cognitive functions including executive functions, attention, working memory, memory, and inhibition to control the train of thoughts, and to monitor semantic and linguistic elements of the discourse. It also involves social cognition to understand others' intentions and feelings (Ybarra, 2012; Ybarra et al., 2008). Quantifying atypical topic variations in prodromal Alzheimer's disease represents an important, and yet under-examined area that may reveal underlying cognitive processes of patients with MCI.

1.2 Topic Segmentation

A key problem in our conversation analysis is dividing the consecutive utterances into segments that are topically coherent. This is a prerequisite step for our higher-level analysis of conversations involving representation of entire conversation by a set of quantifiable measures. Topic segmentation methods first segment the sequence of utterances into a set of finite topics, representing utterances as vectors in a semantic space. Next, they measure the correlation between two adjacent encoded utterances, and finally predict the topic boundary according to a pre-specified threshold value compared to calculated correlations. Based upon the criteria they adopt for quantifying the cohesion among a pair of consecutive utterances, they can be broadly categorized into two models. Assuming the topic shifting is strongly correlated to the term shifting, *lexicon cohesion models* rely on similar terms of each utterance; that is, topically coherent utterances share some common terms within a short window of spoken words. They are learned in an unsupervised fashion and do not require labeled data. Widely used algorithms such as *TextTiling* (Hearst, 1997) and *LCSeg* (Galley et al., 2003) are examples of lexical based methods for topic segmentation. In contrast to lexical based methods, *contextual cohesion models* exploit the semantic knowledge from the entire utterance rather than key terms. These context-dependent models assume that utterances with a similar semantic distribution share the same topic. More recent methods leverage the deep architectures, such as recurrent neural networks (RNNs) (Sehikh et al., 2017) and convolutional neural networks (CNNs) (Wang et al., 2016) to semantically encode the utterance into a vector space. Treating the topic segmentation as a sequence labeling problem, labels (i.e., topics) are then assigned to every utterance. Context dependent models assume that, if two documents share the same topic, the word distribution of these two should also be similar. Despite the potential benefits of extracting the knowledge from the content, there exist several barriers to taking advantage of them in clinical conversations. Successful deep architectures are trained on large amounts of training examples, typically obtained from structured written text such as medical textbooks or Wikipedia. These models perform well in highly structured data; however, their performance degrades once used in unstructured samples, such as social conversations, due to mismatch between the characteristics of testing and training examples. Topic segmentation in conversational text is more challenging than the written text as it is less structured and typically include shorter utterances (e.g., acknowledgements) and disfluencies (e.g., "um" and "hmm").

2   Data collection and participants

For this preliminary work, we used a collection of semi-structured conversations collected randomized controlled clinical trial entitled *I-CONECT* (https://www.i-conect.org/; ClincialTrials.gov: NCT02871921) conducted at Oregon health Science University (OHSU), University of Michigan, and Wayne State University. In *I-CONECT* study, participants engage in a 30-minute video chat 4 times per week for 6 months (experiment group) followed by 2 times per week for an additional 6 months (control group). Conversations are semi-structured, in which participants freely talk about a predefined topic such as leisure time, science, etc. with trained interviewers. Interviewers were asked to engage participants into a conversation by showing picture prompts, share facts, and ask questions related to predefined topics such as leisure time and science. Interviewers were also instructed to minimally contribute to the conversation (less than 30% of total conversation time) and let participants freely talk about daily selected topics. Our analysis includes a total of 45 older adults, 23 with MCI and 22 healthy controls. Table 1 reports their baseline characteristics. Upon completion of Montreal Cognitive Assessment (MoCA) (Nasreddine et al., 2005), a cognitive screening tool to identify MCI, the test results were evaluated at consensus meeting to clinically determine MCI or normal (i.e., clinicians' consensus based-determination).

Variable	Intact n=22	MCI n=23
Age	80.82 (4.87)	84.06 (5.43)
Gender (% Women)	86.36%	68.22%
Years of Education	16.05 (2.70)	15.17 (2.85)
MoCA	26.14 (2.46)	22.00 (2.84)

Table 1: Baseline characteristics of MCI and cognitively intact participants. Montreal Cognitive Assessment (MoCA) score, ranged from 0 to 30, is used as a screening tool and it is lower in MCI subjects.

3 Methods

In our recent study, we presented a method for automatically identifying individuals with MCI based on the count of individuals' spoken words taken from the semi-structured conversations between interviewers and participating older adults (H Dodge et al., 2015; Asgari et al., 2017). We showed that individuals with MCI talk more than healthy controls in these conversations (H Dodge et al., 2015), as they may need to substitute words in the conversation to convey their thoughts. Also, we showed that their lexical pattern, obtained by counting the frequency of words picked from a particular word category such as *verbs* and *fillers*, is different from healthy controls (Asgari et al., 2017). The main limitation of our prior works on linguistic analysis of conversations is ignoring sentence structure and other contextual information relying entirely on word-level features. Enhancing our automatic analysis of clinical conversation, we aim to characterize the relationship among the sequence of sentences, presented in the course of conversation, in order to track the exchanged topics. Our central hypothesis in this work is that patients with MCI may have subtle difficulties with executive and self-monitoring conversation consistency relative to those with normal cognition resulting in more disruptive pattern of exchanged topics within the conversation.

3.1 Utterance Representation

Given the limited amounts of text data in this study, it is difficult to employ deep architectures for learning semantic models. Instead, we adopt LCseg (Galley et al., 2003) algorithm to divide utterances into semantically related clusters. LCseg uses word repetitions to build lexical chains that are consequently used to identify and weight the key terms. A lexical chain is a set of semantically related words inside a window of utterances that

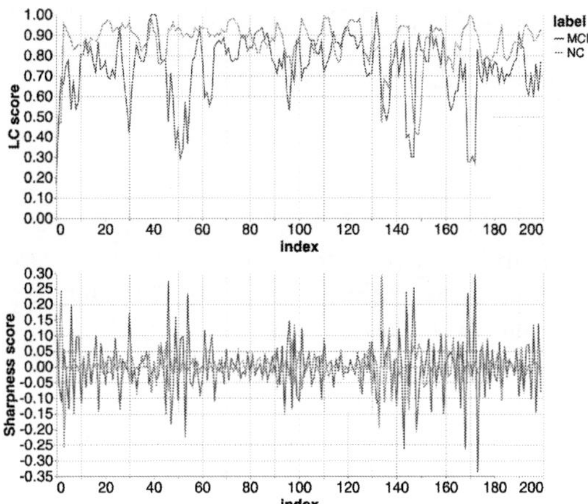

Figure 1: LC (top) and sharpness (bottom) scores of two MCI and NC subjects as a function of utterance index.

capture the lexical cohesion followed within the window. From the lexical chains, it then computes lexical cohesion (LC) score among two adjacent analysis windows utterances.

To predict a topic boundary, LCseg tracks the fluctuation of LC scores and estimates an occurrence of a topic change according to a sharpness measure calculated on surrounding left and right neighbors of the ith center window as :

$$S_i = \frac{1}{2}[LC_{i-1} + LC_{i+1} - 2 * LC_i] \quad (1)$$

Assuming that sharp changes in sharpness score co-occur with a change in the topic, LCseg locates the topic boundaries where the sharpness score exceeds a pre-specified threshold value. LCseg was originally designed to analyze transcription of multiparty oral meetings that typically include six to eight participants. Similar to our semi-structured conversations, ungrammatical sentences are common in such meetings.

3.2 Automatic Measures of Conversation

The top plot in Figure (1) depicts the lexical cohesion scores calculated across the sequence of utterances chopped from conversation recordings of two MCI and normal control (NC) participants. The horizontal axis represents the utterance index that spans from the beginning to the end of the conversation, and the vertical axis represents the lexical cohesion score. As it is seen in these plots, the LC scores of the normal control (NC) participant are smoother with less frequent sharp changes com-

model	ROC AUC	Sensitivity	Specificity	Accuracy
SVM	83.82% (13.39%)	80.77% (19.57%)	77.36% (18.25%)	79.15% (12.44%)

Table 2: Classification results (with standard deviations) for distinguishing 23 MCI from 22 normal controls.

pared to participants with MCI, suggesting a structural difference in the pattern of their discussed topics across the conversation. To measure the variations of the LC score across the utterances, we use Shannon's entropy, an appropriate metric to measure the level of organization in random variables (Renevey and Drygajlo, 2001) and measure the entropy of harmonic coefficients. The bottom plot in Figure (1) depicts the sharpness score calculated on LC score of two MCI and NC participants (top plot) according to Equation 1. The more frequent and yet abrupt changes in sharpness score of MCI subject indicates the higher likelihood of topical changes in the sequence of utterance compare to the NC subject. To capture the frequency of these changes, we adopt the zero-crossing rate (ZCR), a measure that quantifies the number of times a signal crosses the zero line within a window of the signal. ZCR is a common measure in speech processing algorithms for differentiating speech from noise segments (Bachu et al., 2010). Prior to compute the ZCR, we normalize the sharpness score such that it becomes a zero-mean signal. Dividing the entire signal into finite number of fixed-length windows, we compute the ZCR for every window and ultimately summarize the computed ZCRs across the entire conversation using mean and summation statistical functions.

4 Experiments

4.1 Pre-processing and Feature Extraction

Removing the interviewer's speech, we narrow our focus on the analysis of the participant's side of the conversation. For pre-processing of the transcriptions (e.g, removing the punctuation), we adopt an open-source library, SpaCy (Honnibal and Montani, 2017), with its default settings. We also set the minimum number of words per utterance to three words and exclude the shorter utterances. We also trimmed out fillers (e.g., "hmm", "mm-hmm", and "you know") from the transcriptions. Pre-processed transcription of conversations are then fed into LC-seg algorithm where from its output, LC score, we compute the sharpness score. Next, we calculate the entropy of the LC score as well as ZCR of both LC score and sharpness score as described at 3.2.

4.2 Results

Representing a conversation using four measures selected by RFECV (sum and mean of ZCR on LC score, the entropy of LC score as well as the sum and mean of ZCR on sharpness scores), we trained a linear support vector machine (SVM) classifier from the open-source Scikit-learn toolkit (Pedregosa et al., 2011) to validate the utility of proposed conversation measures in distinguishing MCI from NC participants. We used cross-validation (CV) techniques in which the train and test sets are rotated over the entire data set. We shuffle the data and repeat 5-fold cross-validation 100 times. Our results, reported in Table (2), present the mean and standard deviation of four classification metrics: 1) sensitivity, 2) specificity, 3) area under the curve of receiver operating characteristics (AUC ROC), and 4) classification accuracy. Our results indicates that our proposed measures are useful in detecting subjects with MCI.

5 Conclusion

In our clinically oriented study, conversations between the interviewer and the participant provide an opportunity to analyze potential differences in the conversational output of persons with MCI and cognitively intact adults. With the aim of gaining insight about the underlying cognitive processing among patients with MCI, we proposed a computational approach to capture atypical variations observed in the sequence of topics discussed throughout the course of conversation. Our method represents the entire conversation with a set of quantifiable measures that are useful in early detection of cognitive impairment. Despite this promise, a current important limitation to this approach is that the analysis relies on high-fidelity transcription of the conversations which is labor intensive. Furthermore, when applying this approach in clinical trials or to the general population, one would typically add other potentially predictive features to the classification model such as age, gender, education, and family history of dementia. Future studies will need to examine larger and more diverse populations over time and explore the possible cognitive bases behind the findings of the present study.

Acknowledgments

This work was supported by Oregon Roybal Center for Aging and Technology Pilot Program award P30 AG008017-30 in addition to NIH-NIA Aging awards R01-AG051628, and R01-AG056102.

References

Meysam Asgari, Jeffrey Kaye, and Hiroko Dodge. 2017. Predicting mild cognitive impairment from spontaneous spoken utterances. *Alzheimer's & Dementia: Translational Research & Clinical Interventions*, 3(2):219–228.

RG Bachu, S Kopparthi, B Adapa, and Buket D Barkana. 2010. Voiced/Unvoiced Decision for Speech Signals Based on Zero-Crossing Rate and Energy. In *Advanced Techniques in Computing Sciences and Software Engineering*, pages 279–282. Springer.

Michel Galley, Kathleen McKeown, Eric Fosler-Lussier, and Hongyan Jing. 2003. Discourse Segmentation of Multi-Party Conversation. In *Proceedings of the 41st Annual Meeting on Association for Computational Linguistics-Volume 1*, pages 562–569. Association for Computational Linguistics.

Hiroko H Dodge, Nora Mattek, Mattie Gregor, Molly Bowman, Adriana Seelye, Oscar Ybarra, Meysam Asgari, and Jeffrey A Kaye. 2015. Social Markers of Mild Cognitive Impairment: Proportion of Word Counts in Free Conversational Speech. *Current Alzheimer Research*, 12(6):513–519.

Marti A Hearst. 1997. TextTiling: Segmenting Text into Multi-paragraph Subtopic Passages. *Computational linguistics*, 23(1):33–64.

Ildikó Hoffmann, Dezso Nemeth, Cristina D Dye, Magdolna Pákáski, Tamás Irinyi, and János Kálmán. 2010. Temporal parameters of spontaneous speech in Alzheimer's disease. *International journal of speech-language pathology*, 12(1):29–34.

Matthew Honnibal and Ines Montani. 2017. spaCy 2: Natural language understanding with Bloom embeddings, convolutional neural networks and incremental parsing. *To appear*.

Karmele López-de Ipina, Jordi Solé-Casals, Harkaitz Eguiraun, Jesús B Alonso, Carlos M Travieso, Aitzol Ezeiza, Nora Barroso, Miriam Ecay-Torres, Pablo Martinez-Lage, and Blanca Beitia. 2015. Feature selection for spontaneous speech analysis to aid in Alzheimer's disease diagnosis: A fractal dimension approach. *Computer Speech & Language*, 30(1):43–60.

Ali Khodabakhsh, Fatih Yesil, Ekrem Guner, and Cenk Demiroglu. 2015. Evaluation of linguistic and prosodic features for detection of Alzheimer's disease in Turkish conversational speech. *EURASIP Journal on Audio, Speech, and Music Processing*, 2015(1):9.

Kimberly Diggle Mueller, Rebecca L Koscik, Lyn S Turkstra, Sarah K Riedeman, Asenath LaRue, Lindsay R Clark, Bruce Hermann, Mark A Sager, and Sterling C Johnson. 2016. Connected Language in Late Middle-Aged Adults at Risk for Alzheimer's Disease. *Journal of Alzheimer's Disease*, 54(4):1539–1550.

Ziad S Nasreddine, Natalie A Phillips, Valérie Bédirian, Simon Charbonneau, Victor Whitehead, Isabelle Collin, Jeffrey L Cummings, and Howard Chertkow. 2005. The Montreal Cognitive Assessment, MoCA: A Brief Screening Tool For Mild Cognitive Impairment. *Journal of the American Geriatrics Society*, 53(4):695–699.

F. Pedregosa, G. Varoquaux, A. Gramfort, V. Michel, B. Thirion, O. Grisel, M. Blondel, P. Prettenhofer, R. Weiss, V. Dubourg, J. Vanderplas, A. Passos, D. Cournapeau, M. Brucher, M. Perrot, and E. Duchesnay. 2011. Scikit-learn: Machine learning in Python. *Journal of Machine Learning Research*, 12:2825–2830.

Philippe Renevey and Andrzej Drygajlo. 2001. Entropy Based Voice Activity Detection in Very Noisy Conditions. In *Seventh European Conference on Speech Communication and Technology*.

Brian Roark, Margaret Mitchell, John-Paul Hosom, Kristy Hollingshead, and Jeffrey Kaye. 2011. Spoken Language Derived Measures for Detecting Mild Cognitive Impairment. *IEEE transactions on audio, speech, and language processing*, 19(7):2081–2090.

Imran Sehikh, Dominique Fohr, and Irina Illina. 2017. Topic segmentation in ASR transcripts using bidirectional rnns for change detection. In *2017 IEEE Automatic Speech Recognition and Understanding Workshop (ASRU)*, pages 512–518. IEEE.

Daisaku Shibata, Shoko Wakamiya, Ayae Kinoshita, and Eiji Aramaki. 2016. Detecting Japanese Patients with Alzheimer's Disease based on Word Category Frequencies. In *Proceedings of the Clinical Natural Language Processing Workshop (ClinicalNLP)*, pages 78–85.

Liang Wang, Sujian Li, Xinyan Xiao, and Yajuan Lyu. 2016. Topic Segmentation of Web Documents with Automatic Cue Phrase Identification and BLSTM-CNN. In *Natural Language Understanding and Intelligent Applications*, pages 177–188. Springer.

Oscar Ybarra. 2012. On-line Social Interactions and Executive Functions. *Frontiers in human neuroscience*, 6:75.

Oscar Ybarra, Eugene Burnstein, Piotr Winkielman, Matthew C Keller, Melvin Manis, Emily Chan, and Joel Rodriguez. 2008. Mental Exercising Through Simple Socializing: Social Interaction Promotes General Cognitive Functioning. *Personality and Social Psychology Bulletin*, 34(2):248–259.

Author Index